Additional Praise for *Data Driven Decisions*

Joshua Jahani presents a powerful set of tools driven by a pragmatic engineering approach. I appreciate the definitiveness of this work and believe it can be a valuable tool for people seeking to understand corporate value and intangible assets.

—Turk Al Joaib, *Managing Partner,*
Rua Growth Fund, Riyadh, Saudi Arabia

Data Driven Decisions captures the missing part of corporate valuations, driven by the opacity of intangible assets. All financial professionals would benefit from this book.

—Andrei Ugarov, *Former PwC Corporate*
Finance and Valuations Partner, Abu Dhabi, UAE

Data Driven Decisions shows a commitment to the Systems Engineering tools and principles that have helped the field become more popular than ever before. I am particularly impressed with how the methodologies of this book build on an existing body of systems material taught across the world.

—Wesley A Hewett, *INCOSE Chapter President, New York, USA*

The organization of *Data Driven Decisions* serves Joshua's purpose to identify better ways to measure intangibles in the global economy. I believe capital markets benefit from this kind of organization.

—Andrew Hulsh, *Partner and member of the*
private equity practice at Mintz Levin, a leading
international law firm, New York, USA

Data Driven Decisions

Data Driven Decisions

Systems Engineering to Understand Corporate Value and Intangible Assets

Joshua Jahani

WILEY

Registered Offices
John Wiley & Sons, Inc., 111 River Street, Hoboken, NJ 07030, USA
John Wiley & Sons Ltd, The Atrium, Southern Gate, Chichester, West Sussex, PO19 8SQ, UK
John Wiley & Sons Singapore Pte. Ltd, 134 Jurong Gateway Road, #04-307H, Singapore 600134

Editorial Office
The Atrium, Southern Gate, Chichester, West Sussex, PO19 8SQ, UK

For details of our global editorial offices, customer services, and more information about Wiley products visit us at www.wiley.com.

Wiley also publishes its books in a variety of electronic formats and by print-on-demand. Some content that appears in standard print versions of this book may not be available in other formats.

Library of Congress Cataloging-in-Publication Data Is Available:

ISBN 9781394202331 (Cloth)
ISBN 9781394200054 (ePDF)
ISBN 9781394200061 (ePub)

Cover Design: Wiley
Cover Image: © Xuanyu Han/Getty Images

Set in 11.5/14 pt STIX Two Text by Straive, Chennai, India

SKY10055379_091523

Contents

Foreword

By David R. Schneider

PROFESSOR OF PRACTICE
SYSTEMS ENGINEERING
CORNELL UNIVERSITY

Joshua Jahani's *Data Driven Decisions: Systems Engineering to Understand Corporate Value and Intangible Assets* is a novel work that bridges systems engineering and entrepreneurial business, investing, and corporate valuation. Although sometimes considered quite different, Joshua demonstrates how systems engineering tools, processes, methodology, and philosophies, can translate into entrepreneurial success. It takes an entrepreneur of significant skill and ingenuity to approach the mechanics of business decision-making through the lens of systems engineering, and I applaud Mr. Jahani for this work that deeply enriches both fields of practice.

I welcome any endeavor that widens the audience for the methodologies that I have spent decades studying, and this book can only mean good things for the wider acceptance and understanding of systems engineering as it applies to bringing clarity to and reducing the complexity of any given scenario.

David R. Schneider, PhD

***

David R. Schneider graduated from Rensselaer Polytechnic Institute in chemical engineering in 1999, attended Columbia University Film MFA Program in 2001, and earned his master's and PhD from Cornell University in mechanical engineering with a concentration in controls and dynamics in 2007. David has taught at both Columbia University, where he was the highest student-rated instructor in the College of Engineering, and at Cornell University, where he is now the Director of MEng Studies for Systems Engineering, the largest MEng program at Cornell.

With a strong focus on education, David created the first experience in the world recognized by the systems engineering professional society INCOSE as a knowledge exam equivalent, and the only person to have created two experiences earning this honor. Additionally, David created and runs the systems engineering courses for Lockheed Martin's largest Engineering Leadership Development Program. David's main course, Model Based Systems Engineering, is also now officially sponsored by Boeing. David has also received multiple recognitions for his educational work from the Obama White House Office of Science and Technology Policy.

Prologue

Because this is a systems engineering book, I thought it prudent to provide a short historic roadmap first to give you an understanding of my background in relation to the subject matter. So here is a little bit about how this journey all began.

My higher education life began with premedical studies. That was actually the name of the degree, but it looked like a biology course. I graduated with my Bachelor of Science degree in that field, and I found my way to Cornell through a somewhat unusual path.

In 2010, I was accepted to Cornell for the first time to attend their graduate school program in the field of regional planning. I was actually put on a waiting list for the program, and I had applied to the regional planning program because my minor for my bachelor's degree was in political science.

I had really great political science teachers, all of my friends were in political science, and I just emotionally connected with the political side subject matter much more than the science subject matter, which was very memorization oriented and was very much preparing you for medical school.

I decided not to go to medical school, and I decided to apply to this program in regional planning. I did not apply to many. I applied to a couple of schools in New York City, and I applied to Cornell. I visited the program in person during the open house and was able to secure a spot.

My first year at Cornell was actually matriculated as a master's student in the regional planning program, which is not engineering, and the AAP (architecture, art, and planning program), was very public policy focused.

I rapidly realized that this subject matter was not for me. The whole thing seemed too focused on Marxism.

Instead, I was able to use some of the skills I had in mathematics and science that I developed through my undergraduate degree, and I applied to join the program in systems engineering because I genuinely saw a very big overlap between the systems engineering program and my education in biology and premed studies as well as my year of regional planning at the Ivy League university.

So, I matriculated into systems engineering, and I started that program. They only admitted me based on the contingency that I passed some advanced math courses, which I was able to do, and I achieved the grades necessary to matriculate into the systems program.

I then spent a year and a half taking systems engineering courses, so my educational background in the area of systems engineering is quite broad. Most systems engineers received their bachelor's degree in some kind of engineering program, such as electrical engineering. Systems engineering is a very popular follow-up to electrical engineering. Mechanical engineering, physics, and aerospace engineering are all very common precursors to the master's in systems engineering program at Cornell, but I came in through premed, a bunch of Cornell math classes, and a year of Marxist training under the school of architecture art and planning for the regional planning program.

For most of the students who study systems engineering, it is their first exposure to a less mathematical or less technical body of knowledge compared to their undergrad course, but for me, it was the opposite. I came in with a very broad, very nonmathematical background that then got pushed into this funnel of systems engineering to then start applying quantitative or technical metrics to this very qualitative material I had learned. Regional planning is, of course, super qualitative but even biology is fairly qualitative; even though you are studying nature, you are mostly memorizing the effects that nature has on biology, but it is a very macro view of the world.

When I showed up to the engineering program, I was really passionate and really impressed with how I could take these relatively simple tools – which we will get into throughout this book – and then apply them to the more macro frameworks that I understood.

Common tools within systems engineering include devices such as decision trees, which are taught at MBA programs all across the country, or any kind of management science program. But the systems engineering program takes decision trees to a new level.

There is also a failure mode effect analysis, which is essentially used to identify how an assembly line may fail and then to identify probabilities that each of these different failure scenarios may occur and then start to create solutions for them.

As a very qualitative and macro thinker, I was very excited to get exposed to these tools. It made me a very effective student. I ended up graduating with my master's degree, so my educational qualifications include a master's in engineering and systems engineering, and I was able to use this material to develop even more credibility, which is why Cornell ultimately invited me back to its faculty and my current title at Cornell is visiting lecturer.

After leaving Cornell, I went to Deloitte, where I did technology consulting. There you can see a very professionalized version of systems engineering with big healthcare systems or any kind of big technology-system implementation.

After Deloitte, I entered the world of corporate finance and investment banking. I then opened my own consulting firm in the field of investment banking where we rely heavily on systems engineering tools both to run the business and to provide highly qualified advice to our clients. I knew I had found my niche, as we made millions of dollars in revenue in our first year and systems engineering has been a core part of that success and everything that we do.

I now teach several courses at Cornell, which are all very practically focused. The first one is a project where I lead students on how to use systems engineering to perform corporate valuations. The second course I teach is an introduction to systems engineering course that is available to seniors in the undergraduate program and master students in the graduate program.

Jahani and Associates came from the premise that we could provide better growth advisory services to clients than that available on the market at the time. We are a management consulting firm that has an investment banking capacity, but our clients leverage us for growth, so everything we do has to do with expanding a company one way or another. We broadly define the categories for expansion as investment banking advisory services, which include capital placement mergers and acquisitions (M&A), which ties heavily into corporate valuations and corporate expansion, with the second category of growth services we discuss called market expansion.

Everything we do at Jahani is cross-border in one way or another. We will work with a Canadian firm, and we will help that firm expand to Dubai. We will work with a Dubai-based firm, and we will help that company expand to Jakarta. We will work with a Brazilian-based company, and we will help it expand to Saudi Arabia, and we will work with the Saudi Arabian company, and help it do business in Florida or in California. Those are all examples of different transactions and clients that we have actually worked with. The expansion can include investment banking, as I said, but it is bigger than that. It can include something as simple as opening up an office. Jahani and Associates does a lot of joint venture work, and we are the bridge that facilitates those joint ventures for our client companies.

When you are applying these growth services to these very different markets that have very different cultural, financial, business systems, and so on, you have to mesh them together so that they can grow. In every instance, there is a heavy focus on optimization of how the two systems that you are working within a cross-border capacity must integrate.

You do not have that concern if you are only doing domestic deals, which is what most of our competitors and most of the companies in this space do. If a company from Connecticut is buying a company in Pennsylvania, the cultural, financial, and logistics systems are already integrated. All of those basic business processes overlap very nicely. But when you do it in a cross-border capacity, particularly in the regions that we are talking about, they do not interface, so we have to find a way for them to interface. We have to develop a map from one platform to another. In software terms, this would be called an Application Programming Interface (API). Essentially, mapping one system's protocols to another enables the two platforms to "speak" with one another. Beyond software, there may be services we need to introduce to facilitate or amend one system or the other (or both) so that they communicate. There is always a way.

We use systems engineering tools and frameworks to facilitate that, and we use these tools to help communicate the benefits of the intangible assets that a company may have so that they can achieve an accurate valuation.

To provide a real-world example, I will use a simple case-study scenario to demonstrate the various action items, strategic moves, and workflows of providing a solution to a client.

One of our US-based client companies purchased out of bankruptcy a software platform in Spain that the Spanish government had funded via grants. The first step was to gain an understanding of which triggers for growth could use and gain advantage.

The first step is to examine exposure. The company has this software technology that now exists in the United States and has ties to Spain. We would begin with regulatory compliance as it was funded through the Spanish government. They lend their money to help improve the Spanish economy so the company needs to make sure that the asset is free of any obligation or regulatory ties – any liens, or similar problems.

We make sure that the asset does not have any inbuilt requirement that would give the King an opportunity to take it back from you or have some right to your generated revenues. Essentially we are checking that the company is able to own all of the intellectual property clearly across international borders. That ownership component is extremely important, particularly in software. You also see this very often with pharmaceutical businesses or biology companies.

These companies are licensing key patents from offices of Harvard or Yale because these big institutions create so many patents and so much intellectual property, they cannot sell it because it was funded with government dollars, so they ended up giving a perpetual license for zero dollars. In this kind of cross-border transaction between the United States and Spain, it is important to make absolutely certain that there is no copyright for the code that was actually funded by the Spanish government, which is therefore owned by the government, and the company does not have its own rights to it.

Assuming that the business is free and clear of all of those liabilities, the next step is a pretty basic commercial process, which is about how people are paying for the commercial value of the solution and how much they want to pay.

If the business is only doing 15,000 euros per month already, what is the number of customers generating those 15,000 euros? Are they primarily in Spain, or are they scattered about the world?

If the software is Linux-based and open source, what are the limitations in selling it, and do the company provide the software for free and charge a subscription for training and support instead? Analyzing its nature will show us where opportunities for growth exist. If the customer base is predominantly situated in Spain and the 20,000 customers are loyal advocates, we can use that brand loyalty to open up opportunities in the United States and other countries. If the base is using it for a particular reason, for example, as an alternative to a Microsoft product that has a less intuitive interface and perhaps costs more money, the low-hanging fruit in the market begins to reveal itself. Is it large companies or small companies that are likely to make this buying decision? Large companies are more likely to go with the incumbent for trust and credibility because they are less concerned with price and more concerned with keeping their jobs by not making a wrong choice. This means that we are now beginning to see a clear path to small businesses, perhaps with an in-house IT staff member to configure it. This probably means that we need to target small to medium businesses with five or more staff who have an in-house IT person who can understand the implementation process and save significant money on software licensing in relation to the company's revenue.

Now that we have established our laser-focused target client, we need to consider buying cycles. How are they likely to consider a decision like this? If the base price for 5–25 users is just 195 dollars per year, it is too small for a meeting with any kind of major decision-maker. Managers have a budget, and they can autonomously make a buying decision like this, so that removes a barrier for us. They have a corporate account or a corporate credit card that they are able to use. They want to acquire solutions that help them solve these very specific problems, so we need to establish what those are and market the solutions to them.

With this in mind, we can see that e-commerce channels suddenly look like a friction-free sales process rather than a traditional sales team approach.

The capital the company would have invested in a sales team can now go to a Google Ads campaign or a channel marketing campaign with existing networks, such as data centers where customers have a suite of software solutions on offer from the vendor.

Now we are firmly establishing what we have and who wants it, we can also begin to gear our search engine optimization (SEO) so that customers will find us organically and be funneled into an optimized free-trial landing page. This is a very inexpensive growth strategy, but we have to do this preliminary analysis to start to understand the tangible and intangible assets of the offering, which then guides us to what kind of market expansion fit applies.

Is it a business-to-business process with a team doing cold calls, or is it a direct-to-consumer play?

Now we can get to work on an Excel formula to begin drawing a picture of our forward-looking expansion plan to the Unites States and beyond.

This is how we use our framework, the systems engineering piece, to identify the product's core-calling, which leads us to our target, who is obviously a technical person, which usually indicates that they are a smart person, so now we know how we are going to speak to them, what language we will use in our sales, and marketing outreach, and so on. This process continues as we triangulate the path to revenue. What worked in Spain would not necessarily work in the United States or elsewhere so this process must be done in context with the expansion geography.

Another Jahani and Associates' client is a Europe-based tech company that allows you to manage cloud architecture. They have two go-to-market strategies. They acquire customers online through basic SEO and then via sales-channel partnerships. We arrived quickly at these primary revenue generators using the same framework that we explored during the Spanish software case study.

Another one of our clients specializes in digital marketing solutions in the United States. We closed a substantial investment for them, and we helped them to form a joint venture in the Middle East with a conglomerate. It is a perfect case study to show the scope of what J&A is doing. It is not just consulting; it is not just capital.

They were based in the United States, and they specifically wanted to grow internationally because they have digital marketing solutions for physical spaces that apply anywhere, but that is a scenario that works very well in places like the Middle East and Asia. So, they hired us. Ten months later, they had a new institution that was then valued at several billion dollars on the cap table. That institution became their largest institutional shareholder. They also have several of their media devices being deployed in the Middle East.

We did another deal with a logistics company based in the GCC. We formed a joint venture with a large family owned conglomerate in Indonesia. This logistics company in the GCC owns a last-mile delivery platform that helps e-commerce shippers optimize shipping costs and routes so that they can either save money or they can pass the lower cost of shipping on to their customers, thereby building customer loyalty. In Jakarta, Indonesia – a country with a population of two hundred million people – last-mile delivery is a huge challenge. Partnering with our client brought demand to local couriers.

Hopefully, this gives a well-rounded walkthrough of what J&A does for companies, thereby providing the foundational elements to better utilize the information and tools in this book.

Chief Executive Officers (CEOs) are the de facto experts in their business, which means they know how to fulfill orders, they know how to reach their customers, and they know how to solve employee or labor problems or shortages.

He or she is the de facto expert in their domain, where they start to fail, and where they leverage any service provider is where there is an activity that is needed that is not core to the business. Think about a logistics company that needs an attorney. This logistics provider grew up in FedEx, and she or he is an expert in all-things-delivery, but there is some kind of dispute, and they need legal representation that is not a core competency of the business. This is where they leverage any kind of service provider.

Where we come in very valuably is helping them think about their business within the different systems and economic frameworks of these regions. For example, Chad Thomas from Connecticut has never even thought about the Middle East or Asia, but he knows that these are large markets that can exponentially generate business growth for him.

Another profile of a founder that we found to work with very successfully is Vishal Irma. Vishal is from India, and he knows that Dubai is an extremely relevant market. He knows that Egypt has over a hundred million people and that countries like Iran consume the most bananas in the Middle East. But Vishal went to Brown, and he got his MBA from MIT. Vishal's parents are familiar with these markets, but he himself does not have any kind of expertise or exposure, so although philosophically, he understands that these markets are important, he does not have any on-the-ground experience, and he could not possibly do it himself. It is almost always cheaper to hire us and pay our consulting fees than to try to do it themselves.

Systems engineering is driven by a set of tools that engineers use to get other engineers to work together. Systems engineering is most common in fields like aerospace, where you have to build a rocket, and it has got to go to the moon. You have thousands of engineers all doing little pieces of a massive design. Because systems engineering has this collaborative and highly quantitative nature around processes, it lends itself to analyzing intangibles.

Intangibles can be their own massive separate domain, but it is important to scope intangible assets within the context of systems engineering and valuations.

With systems engineering, it is about what you can conceivably measure. For example, you have a musician, and the musician is phenomenal and world-renowned. An intangible asset of this musician could be the number of views or the number of clicks, or the number of followers that they have on a social media platform. Certainly, sales of albums are relevant, and that is directly correlated to revenue, so that would be considered a tangible asset since cash is overall tangible. But if you have two musicians, both of whom are relatively identical in sound and style, and one has ten million YouTube views, and the other one has one thousand, but neither has monetized those views, one of those is still more valuable than the other, regardless of the money that has been made.

That is how intangible assets matter within the context of a non-financial situation. Systems engineering would allow you to diagram what kind of digital activity the musician with ten million views has done and how their digital marketing strategy has created ten million views versus the musician who has one thousand views and sounds exactly the same.

Linking these concepts of intangible assets with systems engineering tools is a core focus of this book because it has a direct impact on valuations, as we will explore in Part I.

There are three major variables that are inside this multivariate equation of this book. The first one is systems engineering – the tools, the principles, and the techniques that matter within the context of this subject. The second element is corporate valuations, which is ultimately what a company or an asset is worth. We will use the term "corporate valuations" relatively loosely here.

In the example I used with the musicians, they are not businesses themselves, but you can still value both of the profiles of those two musicians.

The third variable in the equation that this book is seeking to analyze is intangible assets, which are generally identified as technology-driven, off-balance-sheet assets that can impact (either increase or decrease) the valuation of a firm.

To summarize, before we get into the guts of the principles, this book is about corporate valuations, systems engineering techniques, and principles for understanding intangible assets. When you combine corporate valuations and systems engineering, it keeps you focused on a very specific world of capital markets – which is about how much a company is worth and what impacts that worth up or down. Then, systems engineering, which is this platform specifically dedicated to keeping engineers – which can include operations research engineers – all on the same track and speaking the same language.

My premise is that systems engineering employs very specific techniques and principles that allow us to understand intangible assets, specifically within the context of corporate valuation. What is fundamentally important to you in deriving value from this book is that there are three variables that are constantly iterating and evolving as we go through this text. The first one is corporate valuations and, specifically, what changes a corporate valuation. The second is intangible assets.

Those are broadly defined and will be defined later in the book by anything that is off the balance sheet, and you cannot touch, mostly driven by technology. Then, the third variable is systems engineering which is the tactile and utile tools that we can use to understand the first two variables in greater detail.

It is important that stay focused on these three variables and constantly come back to them at any time when the subject matter feels like it may be getting too detailed or off course for you because we have gone very deep into a particular element of one of the three variants.

Introduction

Systems Engineering Overview: Customer Affinity and Context Diagram

This book is primarily concerned with corporate valuations and systems engineering in relation to how they can be combined to help business owners, analysts, investors, and advisors understand the intangible assets of a business.

To achieve an adequate examination of the three issues of corporate valuations, systems engineering, and intangible assets, we have to break each of them down into their components to understand exactly what we mean by the respective terms.

Corporate valuation is a simple term to understand; it simply means the value of a company. Valuation, as we will get into later in this book, is generally determined by one of the two methodologies – the income methodology and the comparable methodology. The income methodology is derived from the forecast of the cash flow of the business that is then discounted back based on the time value of money. One arrives at the present value of the business based on its cash flows. The second method of corporate valuation that is commonly used is the comparable methodology, where one locates similar companies that have gone through transactions in capital markets based on some kind of multiple or based on some kind of identifier and then applies that identifier to the business in question.

Multiples are most commonly the enterprise value of revenue or enterprise value over earnings before interest, taxes, depreciation, and amortization (EBITDA), but identifiers in multiples can include something as simple as product development or a sophisticated management team.

The second element is systems engineering, which we will discuss in more detail later in this chapter. Systems engineering, which is highly quantitative, is a tool generally available in graduate-level courses to help engineers work with other engineers. It is driven by a series of tools, which we will find very useful in the discovery of corporate valuations and intangible assets.

In the world of consulting and operational tools, the last part of the triangle is intangible assets, which, fundamentally, are assets that you cannot touch or feel, like know-how, processes, expertise, software, patents, trademarks, copyrights, workflows, and a variety of other things that we will get into.

What I would like to focus on right now are corporate valuations, systems engineering, and intangible assets. All three of these concepts converge or overlap within their academic and practical domains. Systems engineering is likely the most niche or specific of the three domains, and therefore, we will investigate it first to determine the relevant techniques and principles that pertain to this book.

The best way to describe systems engineering as a practice is through the concept of a T-shaped engineer. All systems engineering programs across the world identify and discuss this idea of a T-shaped engineer. A T-shaped engineer is someone who is broad in all categories, that is, a person having in-depth knowledge in a specific domain but at the same time capable of working across assorted domains. In a nutshell, they have cross-disciplined expertise.

They have, for example, familiarity with aerospace, mechanical, electrical, operational, and industrial domains but have a single deep discipline in one area, for example, mechanical.

The vertical shaft on the T in the metaphor indicates that they are very deep in a single industry or single subject-matter domain, and the horizontal bar on the T refers to their being experienced with, or their capacity to work across, a variety of industries. Within the context of corporate valuations and intangibles, we need to be more specific with this T-shaped terminology.

As for intangibles, they are applicable to every single domain that I just mentioned in my T-shaped engineer example – aerospace, mechanical, and so on. Intangibles in and of themselves do not constitute the depth of engineering. Except, maybe, in computer science, where the discussion is singularly about software code. But that would be a pedantic definition and would be too small for the scope of this book.

The deep expertise in one discipline that applies to the systems engineering principles of this work is in the domain of corporate valuations, and corporate valuations – or corporate growth – and is a very specific, and in some ways, a T-shaped domain of itself that draws on strategy macroeconomic and microeconomic business strengths and unique business factors that help identify why one company may be worth a different amount than another.

It is largely driven by financial aspects. If you look at corporate valuations, the major focus is on the financials of the business. But the financials are not the only way to value a business, particularly in the age of intangible assets, which we will discuss in other parts of this book. The bottom line is that systems engineering is a tool in the domain of corporate development and corporate valuations. It is a set of tools that engineers use to work with other types of engineers.

The techniques and principles of systems engineering include fault mode effect analysis and process flows with an analytical hierarchy. It is all about being able to quantify the decisions that are made (intangible systems), specifically with an engineering mind-set, which is always more quantitative and more focused than what one may be familiar with in business school.

For example, a common tool of systems engineering can include a customer affinity process, which is a mining technique that aids, or enables, the process of breaking down customer comments or customer feedback into several important attributes or categories. This is most applicable for work zones such as product development. The first step of processing customer affinity (turning buyers into brand loyalists) is to combine hundreds of pieces of customer feedback, that is, hundreds of comments, into specific categories to establish sentiment across these verticals.

In the example of a mechanical system, it could be about a certain speed, a certain level of durability, or a certain level of physical features that customers are looking for. For example, a customer who needs a hammer may give comments such as, "I want my hammer to be eight to fourteen inches long." As these comments are grouped, it allows an engineer, not a businessperson, to know what the most relevant pieces of information are so that they can inform other engineers.

The deliverable of a customer affinity process is not a vastly complicated calculation. It is itself a very simple business artifact that consultants and financial professionals are accustomed to saying over and over again. What makes it unique within the context of this book and within the context of engineering is that it is a fundamentally quantitative way to identify, one by one, all of the different customer comments and categorize or synthesize them into sets that the customer or the other engineers may need to know.

Categories can include any kind of attribute that is a relative grouping hierarchy of the system in question, such as safety and durability.

Speed may be a very good category for an intangible system that is mostly software based. Another example of a systems engineering tool that will be very much applicable to customer value corporate valuations and intangible assets are context diagrams and context

matrices, which involve brainstorming and thinking of components that may interact with the system. This can involve external
stakeholders, internal stakeholders, and people that are familiar or
unfamiliar with the system.

These systems engineering principles can be explained with the
help of the context diagram in Figure 0.1. You take a box, and you
write a verb about how one box interacts with the system.

For example, when a user logs into a software platform, a business engages its customers, customers engage the business, and so
on and so forth.

The number of these external stakeholders and the way in
which these verbs are articulated indicate how the internal or external stakeholders interact with the system, which in our example is
the business.

It is a highly measurable, highly specific way to understand how
the stakeholder will interact with the business, and this is a thought
process that is enabled by a context diagram that has been highly
popularized by these physical and mechanical systems that we see
inside systems engineering.

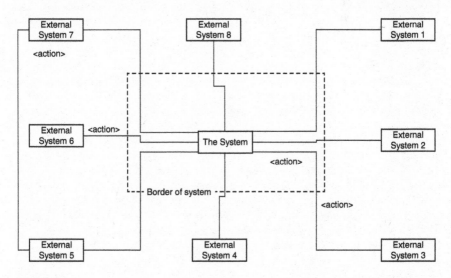

Figure 0.1

We will go through ten to fifteen of these examples as we talk about different systems engineering tools and how they can relate to corporate valuations.

To drive this point home specifically: context diagrams are a very effective way to understand how internal and external stakeholders are interacting with the business and the actionable impacts they are having on the business. Auditors analyze and criticize a business, customers buy from the business, customers return to the business, and customers promote the business. With a customer affinity process tool, it is about grouping customer feedback into consolidated classified sets so that executives and investors can make decisions about which customer requirements are related to each other.

Without the use of systems engineering, these tools largely get relegated to random ad hoc business units, often in sales and marketing, which I will argue is the wrong place for them to be and ultimately makes valuation less accurate because these business functions do not understand these uses.

Part I

Intangible Assets

Chapter 1
When Intangibles Matter Most
Defining Intangible Assets

I n order for us to understand when intangibles matter, we need to define what intangibles and intangible assets are, specifically within the context of corporate expansion and valuation.

Broadly defined, intangible assets are assets that you cannot touch and feel. This includes but is not limited to intellectual property, copyrights, trademarks, software code, movies, written works, and visual arts. Intangibles that cannot be touched are elusive and difficult to define – simply given the lack of their physical properties and characteristics.

Within the context of intangible assets specifically, we can break these down into two major categories – identifiable intangible assets and unidentified intangible assets. This categorization is derived from accounting standards, generally accepted accounting principles (GAAP), and International Financial Reporting Standards (IFRS) methodologies that are largely used in financial reporting accounting and large publicly traded companies.

Identifiable intangibles are those intangible assets that cannot be touched or directly seen with the naked eye. But one can point to and meet a separate set of criteria – namely, they must be separable, measurable, and identifiable. We will come back to these sub-criteria in due course.

The second category of intangible assets is unidentifiable, which are commonly referred to as a "plug."

In the world of accounting, sometimes you know there is value, but you are unable to point to exactly what that value is – however, you paid for it anyway.

Identifiable and tangible assets are the most important components of corporate valuation because they meet traditional reporting standards. As technology advances, our ability to measure things grows with it. In the world of advertising technology, for example, the amount of measurable engagement has increased significantly since the dot com boom of 1995.

The three new criteria for defining an identifiable intangible asset are:

- Separable
- Measurable
- Identifiable

To understand how an intangible asset, which is identifiable, can be separable, measurable, and identifiable, let us see an anecdotal and humorous example.

Consider a management company that has incredibly good-looking management staff members. Assuming there has been some kind of third-party study and it has been independently verified that all the management staff members are attractive to other members of the human species. That may be measurable – they are more attractive than other members of the human race. But it is not separable. For something to be separable, it has to be licensable. It has to be something you can extract from its core source and then sell, license, or transfer as its own unique "separate" identity. Separating someone's attractiveness from their management is impossible. These two items, which arguably are intangible assets, high-quality management, and high attractiveness, cannot be fully separated from each other. Therefore this example, which is meant to be humorous, fails the test of separable, identifiable, intangible assets. We would hence reject it from the identifiable intangible asset category, and we would apply it to the unidentifiable category.

Now consider, for example, an identifiable intangible asset such as a patent. A patent is a written equation, a written recipe, or know-how that allows utilization to accomplish a stated objective. Patents can be licensed. Patents can be sold as a separate consolidated set of rights that allow the owners or the licensees of the patents to perform some kind of function that has been protected by US law. Therefore, patents are separable. But are patents identifiable? Yes! A patent is something that you can identify as describing a unique set of processes, know-how, or ingredients that accomplishes some kind of unique objective when combined. Is a patent measurable? Yes! You could have one patent – you could have twenty patents. You could have patents that vary in complexity. Therefore, a patent as an identifiable intangible asset meets all three criteria of being separable, measurable, and identifiable.

Let us consider another example of an identifiable intangible asset that is less common or less obvious and can include software code.

Software code is lines of copyrighted language that, when placed into a program, accomplish some set of functions as defined by the parameters of the code itself. Is it separable? Yes. It is separable because you can extract this software code. You can cut it; you can copy it, you can paste it, and you can send it to your friend in a letter. Is it identifiable? Certainly: You can point to the code and say what this code does specifically. It allows a user or an owner to accomplish some set of functions. Is it measurable? Yes, software code is measurable because you could say that the software code has five lines, five characters, or fifty lines and fifty characters.

Software code can also be protected by copyright. It is a less common or less obvious example of an identifiable intangible asset, but it is identifiable, nonetheless.

Unidentifiable intangible assets will be dealt with in more detail when we get to the examples of mergers and acquisitions. But they really are a catch-all phrase for things that have not been able to currently be defined within the category of identifiable intelligence.

As technology advances and as we learn more about how to measure, separate, and identify intangible assets, the ecosystem or world of unidentifiable intangibles will no doubt become even smaller, and identifiable intangibles will become larger and more relevant in the ecosystem of this space.

These terms are essential to define intangibles within the context of corporate expansion and corporate valuation. Valuation, patents, trademarks, copyrights, and know-how all impact the value of the business. The more a firm is able to measure them, the more they could positively impact the value of the business that owns these assets.

For corporate expansion, these assets are just as, if not more, applicable because a company that has patents and has protection around its intangibles will be more competitive in the corporate expansion category than a company that does not.

We will address this in more detail in the third part of this book. It is extremely important that the reader understands that defining intangible assets is an essential part of the journey to understanding when intangibles matter most. Not all intangibles are created equal.

Identifiable intangibles are more important than unidentifiable intangibles, and the category of an unidentifiable intangible changes in the context of the business in question.

Chapter 2
Examples
of Intangible Assets

Goodwill, Reputation, Brand Recognition, Patents, and Copyrights

In addition to defining intangibles within the context of corporate valuation and expansion, it is important to create very concrete examples of what these identifiable and unidentifiable intangibles are. The most common examples are defined by accounting standards.

We have already discussed two of these in the identifiable and unidentifiable versions, but it also includes goodwill, which is a highly specific accounting term used solely within the context of mergers and acquisitions (M&A). M&A will be explored and further defined a little later in this book, but the important thing for you to understand is that goodwill, as an intangible asset, has a measurable impact on the valuation of the business. Goodwill is nebulous. It is easy to define the amount of goodwill, but goodwill in and of itself is a term that lacks specificity and is often overlooked by accountants, financiers, and investors as a mere mathematical plug.

To be clear, it is a mathematical plug that is defined by a purchase price allocation equation. But to write off goodwill simply as a mathematical plug with no further inherent information or no further value to understand the corporate valuation of a firm is a mistake, and we will address that further presently.

Examples of intangible assets apart from goodwill include anything that meets the criteria of being separable, measurable, and identifiable.

These three characteristics can apply to things such as patents, copyrights, trademarks, brand recognition, software code, reputation, management experience, and the experience of a workforce, as well as many other things that are all specific to a firm and its industry.

It is important to understand that not all intangibles are the same across all firms. A firm specializing in advertising technology will have industry-specific intangible assets such as customer reach and customer data aggregation. A professional services category may have a highly experienced team and a high number of licensing processes, as well as relevant experience that a client or a customer may pay for, and this differs greatly from the intangible assets of the ad-tech company.

These examples are important to always point out. Humans are very good at intuitively identifying which intangibles matter and which examples apply based on the context of a firm or whether we are talking about valuation or expansion, the industry, or a specific goal.

Follow your intuition. When you are able to follow your intuition, you will be able to collect information about intangibles that help you focus on making your intuition more specific so that you can drive toward better business outcomes. That is essentially the point of this book.

I want to enable the reader to use these simple and accessible systems engineering tools to start intuitively pointing at intangible assets.

Chapter 3
The Rise of
Intangible Economies

Intangible Assets Are More Valuable

Intangible assets are more valuable than tangible assets in the twenty-first century. We need to define intangible assets within different scenarios. The phrase intangible assets, as I have already demonstrated, only means an asset that you cannot touch or feel.

The exercise of this book is to make the reader understand that intangible assets matter more than tangible ones – that intangible assets impact corporate valuation and that the definition and identification of intangible assets change based on the scenario, the industry, and the identifiers of a firm.

A discussion of intangible assets requires a discussion of the history and a discussion of financial statements. The work of Baruch Lev, which has been published by Wiley in his book, *The End of Accounting*, is an excellent resource for understanding how intangibles have become so misunderstood in today's financial reporting economy.

Taking an example directly from Baruch, consider a financial statement such as Profit and Loss (P&L). This financial statement includes highly relevant and valuable information: revenue cost of goods sold, gross margin, operating expenses, and net income earnings before tax. You need this information to be an effective manager. This information has not changed in financial statements over the last one hundred years or more, but the world and economies have changed.

Starting in 1995 – with the dot-com boom in the rise of technology – economies have been undergoing a rapid evolution driven by computers. The growth of computers and the growth of technology in everything that human beings do, impact how businesses function and how they are valued. But when you look at a financial statement, this reporting tool has not changed to reflect the impact of this intangible economy that has grown exponentially, particularly since 1995.

It is driven fundamentally and solely by the Internet. The Internet and the connectivity of the human race are changing business and culture so rapidly that financial statements cannot keep up, and they have become completely out of date ever since 1995 or 2000, shortly after the dot-com bust.

This is a new problem – measuring intangible assets. Defining intangible assets was not something people thought about. Defining intangible assets was not considered by investors, managers, or accountants until 1995.

As of the writing of this book, this problem is less than thirty years old. So, it is no surprise that modern financial reporting techniques are behind. But to properly derive value from this book, you need to understand that this is a problem specifically tied to technology. Now, as a result of this lack of reporting on this tech-tethered problem, there is no perfect solution.

In this book, I am recommending systems engineering as a tool that business owners and investors can use to identify these intangibles and study them more closely so that they can positively impact valuation. But certainly, these tools are not a perfect magic bullet or perfect solution for all corporate valuation conundrums created by the rise of technology and the updating of financial statements.

Since 1995, the data have shown us that the value of firms increasingly resides in their intangible assets. This can be seen through a variety of different metrics, including enterprise value over EBITA multiples, as well as just considering the basic balance sheet of an asset or business and what it trades for in the open market.

The book value, balance sheet value, or equity value of a business is always below the market value unless the company is in distress.

Chapter 4
Defining Intangibles

Now let us talk specifically about defining intangibles within different corporate finance scenarios. These can broadly be defined as valuation, expansion, or commercial. The commercial scenario is the most simple and obvious of the three. What does your customer care about? You sell watches. You sell software. You sell chessboards.

Your customer wants a watch that meets their needs, whether it is a luxury, high-brand-value, or an economic need that simply tells the time when they are working a nine-to-five work shift.

Your chessboard customer is looking for either an artisanal brand or something that is functional so that they can teach their children how to play chess. A customer that is looking for software is looking for a computer-based solution that meets very specific needs, which is, of course, dependent on their industry.

That is the commercial impact of intangibles, and that is a commercial scenario. The person that the intangibles matter to, in this example, is your customer.

Defining intangibles within a corporate valuation setting largely falls into the category of investment banking. This can include the buy-side and the sell-side of investment banking. When I use the word investment banking here, I am specifically talking about mergers and acquisitions (M&A) and, in some cases, some kind of capital placement. I will ignore underwriting, and I will ignore sales and trading, which are both very large and highly lucrative areas of investment banking, but I want to push those verticals outside the scope of this book so that we can focus on the intangibles that really matter when driving a change in a business.

There are also market data that allow us to measure the impact of intangibles on M&A and capital placements, which are somewhat different from the market data available in the public markets with underwriting, sales, and trading. (See the Price–Earning (PE) Ratios section later in the book.)

Sellers want more money, sometimes to their own detriment. Sellers want higher valuations, sometimes to their own detriment. Buyers want to pay lower evaluations, sometimes to their own detriment. Therefore, sellers are using intangible assets to communicate the value of their business more effectively.

It is important right now for you to understand that intangible assets are not the key to getting a pre-revenue business to be worth tens of millions or hundreds of millions of dollars. Buyers will place their own value on assets based on their own methods that have nothing to do with intangible assets.

Intangible assets are not a way to inflate a valuation such that it is irrational, but intangible assets are extremely effective methods to communicate the specificity of the value that is inherent to a business, and this is very important.

You should not think that on the sell-side of a corporate valuation exercise, these tools will allow you to use intangible assets to communicate an absurd or unreasonable valuation for a business. That is not the case. No tool can do that, and I highly encourage anyone reading this book to avoid falling into that pitfall.

What these tools will allow you to do is to take the intangible assets that you have – and that no one argues that you have – and communicate their value more effectively so that the business seems more attractive.

There is no linear or directly correlated relationship between intangible assets and reports or these tools and evaluation of a business. There is simply the presence of an understanding of a firm's intangibles and the fact that this understanding automatically makes the business more interesting or attractive at whatever valuation the market has already set.

For the buy-side within corporate valuation, firms have their own methods for buying businesses – private equity-based buys on multiples of base cash flow. Strategic buyers will buy based on some kind of unique business case that they usually finance with their own equity. Once they acquire these assets, they have to be able to adequately report on them. This gets into purchase price allocation, and that is something that we will discuss later in this book.

Buyers are not necessarily always looking to minimize the value of an asset, but buyers need to understand intangible assets in a different way than sellers do. The seller needs to understand their intangible assets to communicate the value to their customer and to communicate it to the buyer. The buyer needs to understand the intangible assets of the business they acquire to make sure they are able to run the business in a way that makes the acquisition accretive. It is a subtle but important difference between the psychology of the seller and the psychology of the buyer.

Lastly – within the different scenarios of intangible assets pertaining to corporate expansion – corporate expansion is about acquiring customers in a region where a firm is not presently represented. It requires the intangible value of a business to be communicated to a different customer demographic. If the customer demographic changes based on culture, region, location, and time zone, then those intangible value statements must also change.

Chapter 5
Valuing Intangible Assets

Strategic versus Intangible Assets

Now that we have identified the difference between intangible assets, when they matter most, in what situations they matter most, and what defines an intangible asset – such as separability, measurability, and identifiability – we can move into a new set of criteria for intangible assets.

We have discussed how to identify what an intangible asset is, and we have discussed how to identify when an intangible asset matters most. We have also explored which scenarios can influence the valuation of intangibles in one instance more than another. What we now need to address is how these variables can come together to affect value. But first, what is value?

Quite simply, value is the dollar amount that is assigned to an asset or to a distinct set of assets that someone will pay for. If a pencil is valued at a million dollars based on its raw materials, based on its scarcity in the market, and based on its demand from investors, then people will pay one million dollars for the pencil.

It is important to realize that valuation is partly intellectual, but it is also a business exercise. For the pencil to be worth one million dollars, the pencil must be purchased for one million dollars. Valuation is not an exercise in a vacuum based on mathematics alone. It is an exercise based on human behavior, economics, and on the market – what amount of money buyers or investors are willing to spend.

The definition of value as a dollar amount assigned to an asset or set of assets is based on mathematical inputs as well as market and economic transactions. So, how can we apply this very basic definition of value to the intangibles that we are analyzing within this book? It comes back to the criteria that we used to identify them, which, again, were separability, measurability, and identifiability.

Let us examine a patent as our first example. A patent is a right granted by the government allowing you to utilize or monetize a proprietary design, process, or statement and lets you take legal action against anyone who makes, uses, sells, or imports your design, process, or statement without your permission.

Patents are:

- **Separable:** you can transfer one patent between three parties that are unrelated.
- **Identifiable:** you can look at a patent and identify that this is, in fact, a patent.
- **Measurable:** you can have one patent; you can have twenty patents.

Then, the question becomes, "What are the mathematical inputs that are applicable to this patent based on how it meets these three criteria?" To determine what the asset is worth, there are three methods of valuation that we will talk about in this book, and I will review all three of them now in an anecdotal manner. More information about valuation is provided later in the book, but this book is by no means meant to be an exercise or a lesson in valuation. That subject is covered in many other books and many other educational programs that I encourage you to pursue if they are of interest.

The first category of valuation is income valuation which is the most common form, and it is fundamentally driven by a mathematical calculation for determining how much cash or income an asset will produce within a certain period in the future. You own a pencil. This pencil pays you a dollar dividend each year; therefore, it will pay you five dollars over the next five years, and you discount the value of one dollar in years two, three, four, and five back based on some kind of discount factor. This is a cash flow equation. This is a discounted cash flow.

This is very common, and this is method one of income valuation for an intangible asset, which in this case is a patent or a pencil.

The way you would identify the income a patent creates is you would say owning this patent allows me to perform a certain process, and by having this patent, I will receive some kind of royalty. There is a relief from the royalty method, where you identify the amount of income a patent generates, you apply a percentage to that income, and that becomes the cash flow that is assigned to the patent.

The second valuation method is asset-based or cost-based accounting. If you build a building and that building has ten million dollars' worth of steel, plus ten million dollars' worth of lumber, plus ten million dollars' worth of furniture, and it costs you one million dollars to receive all the building permits, you can value this asset at ten plus ten, plus ten, plus one, which is thirty-one million dollars. It cost you, the developer in this case, thirty-one million dollars to create it; therefore, as long as you have your receipts, the value of the asset should be comparable to that.

Juxtapose the same example to the patent – a patent may have required you to conduct ten years of different experiments, ten years of trials and errors, and a thousand different failed experiments. You can assign a dollar value to each of these events, and you could say that for someone to recreate this outcome, it would cost them at least the amount of money that you have spent to arrive at the patent.

Asset valuations or cost-based valuations consistently undervalue assets. They are the bare minimum that one would have to spend to build the same functionality or capability that the asset creates. Cost-based or asset-based accounting methods do not take into consideration the amount of time and the amount of trial and error that may exist with one party over another in order to achieve the same results as the successful asset that is the subject of our study right now.

The last methodology that we discuss for valuation within the context of intangible assets is market-based valuations. Market-based valuations are where you have pencils that are freely traded on the market. Pencils cost a dollar. Pencils create three dollars of income in five years, so calculated on their asset-based valuation, they are worth a dollar but based on their income valuation, they are worth a little less than three dollars based on the discount factor. However, because there are only thirty pencils in circulation, and there are a hundred million people who need pencils, suddenly, the value of a pencil is well in excess of the asset- or cost-based valuation as well as the income-based valuation.

This example of pencils is regularly taught in MBA programs, most introductory M&A courses on capital markets, and most introductory business courses on capital markets. It shows the power of supply and demand over price and price elasticity.

Market valuations most often overvalue an asset or rather provide the highest possible value that an asset can command in the open market. The reason I bring up market valuations here is not necessarily because there are patents that are freely traded on an open market that could be applicable to this valuation methodology – but assets that are made up mostly of intangible assets which are traded in open markets are companies.

There are companies that will be trading at many factors over their book value, or – as with the examples of the advertising technologies we talked about within the technology giants – they will pay many multiples more than the book value or the tangible asset value of a business. This is because these assets carry a certain amount of intangible asset value that companies are valuing when they acquire these businesses.

Chapter 6

EBITDA Is Not the Best Way to Value Intangible-Heavy Companies

EBITDA, as mentioned earlier, stands for earnings before interest, tax, depreciation, and amortization. EBITDA is a metric that is derived from an income statement or profit and loss statement and is meant to be a proxy for the cash flow of a business.

In most GAAP and IFRS accounting methods, net income or net profit is subject to deductions that are a result of non-cash expenses.

Depreciation is a non-cash expense that exists on an income statement that lowers a taxable net income but does not actually require money to cover the expense. An example of depreciation is when one buys a large machine – let us say, for the sake of an example, that the machine is worth a hundred dollars, and the machine is straight-line depreciated over ten years at ten dollars a year, that machine would go from a value of one hundred dollars to ninety to eighty, seventy, sixty, and so on dollars per year and in ten years, the machine would be worth zero.

That depreciation expense is recorded on the income statement and deducted from taxable net income to help lower tax burdens for businesses based in the United States and Europe.

Amortization follows the same logic. It is a non-cash expense for intangible assets. We will get into amortization more in this book when we talk about purchase price allocation. If you buy a patent, the patent has a useful life of ten years. You paid a hundred dollars for the patent. The patent is straight-line depreciated over ten years. Therefore, the patent goes from a value of a hundred to ninety, eighty, seventy, sixty, fifty, and so on, until it is worth zero after ten years from when you buy it. That ten-dollar deduction each year would be seen on the income statement.

Interest is added back on the assumption that companies are bought on a cash-free/debt-free basis. This is common in private equity markets and taxes are always subject to the specific company that is buying another, which is why taxes are added back to net income to reach EBITDA.

EBITDA is a proxy for the cash flow of the business. The bottom line, plain and simple, is that EBITDA is cash flow. EBITDA is by far and away the most popular metric to value any business, anywhere in the world, in any economy.

EBITDA multiples are meant to indicate what a business is worth to a buyer based on market comparables, which we have just reviewed.

The problem with EBITDA – when it comes to intangible-heavy companies – is that EBITDA is naturally suppressed by investments in intangible assets. To understand this, we need to remind ourselves what a specific intangible asset is. Intangible assets are identifiable or unidentified. For example, creating the patents, copyrights, trademarks, or goodwill must be valued. For the sake of this example, think about intangible assets as software code. You are company X, and Company X has the most sophisticated software imaginable for diagnosing blood illnesses based on a simple blood test. You are able, as Company X, to deliver and create this diagnosis by utilizing your incredibly vast, sophisticated, and complex software code to analyze the markers of a blood sample and then determine what kinds of diagnostic insights can exist.

To develop and maintain that complex, sophisticated, and world-changing software code, you, as the CEO of Company X, must maintain a staff of software developers, which is expensive. In any market, software development talent carries a premium.

In addition to having a staff of software developers, you have also spent years developing this code through trial and error and years making sure that this code is workable and that it has the right number of interfaces – whatever may be applicable to the example that we are using.

All of those costs for this world-changing algorithm appear on an income statement as operating expenses. They are deducted from revenue each year based on traditional GAAP or IFRS accounting standards. The problem with this, when it comes to an EBITDA valuation, is that, since these expenses and these investments are applied to financial statements through an operating expense that fundamentally suppressed net income and therefore suppresses EBITDA, you cannot reasonably add back the costs for these developers because there are no widely accepted accounting principles that indicate what percentage or what kind of utilization of a software developer for this world-changing algorithm would be appropriate. A business cannot add the costs of developing this asset back to a balance sheet or back to a P&L to claim that an asset has been developed – purely because of the fact that the asset is intangible.

If you juxtapose this example to the acquisition and utilization of a tangible asset, it is a very different scenario. If a company buys a machine for one hundred million dollars, that acquisition of the machine is most often done through an item of capital expenditure.

Capital expenditure (CapEx) does not appear on the P&L. It is a balance transfer on the balance sheet that debits cash and credit assets, and shows that cash has been spent for a one-hundred-million-dollar asset on the firm's balance sheet.

CapEx does not suppress EBITDA for tangible assets – Operational Expenditure (OpEx) suppresses EBITDA for intangible assets. So therefore, a multiple based on EBITDA in an intangible-heavy business would fundamentally be suppressed or reduced because EBITDA itself is suppressed or reduced because all of the costs to develop the intangible assets are being captured on the financial

statement as an operating expense. This is a very important fundamental of the challenge associated with valuing intangible assets.

The accounting rules for valuing intangibles are very different
and non-standard when compared to typical accounting principles
that we use for tangible assets. This is proven in the market. EBITDA
multiples for tech-heavy businesses are higher than EBITDA multiples for manufacturing-based businesses or EBITDA multiples
for tangible-heavy businesses. That is because the EBITDA is suppressed artificially for an intangible business so the multiple must
be higher to account for the proper value that the market wants to
assign to the asset.

EBITDA is not the best way to value intangible-heavy businesses, but for better or worse, it is the primary method that we
use, and by and large, in capital markets both private and public,
it is the only widely accepted method. But there is a problem. The
problem is defined through the way operating expenses and capital expenditures are accounted for. Some intangible asset expenses
can be "capitalized" and added back to the balance sheet, but the
rules for doing this are very "custom," and everyone follows them
differently; they are almost never 100 percent. Because EBITDA is
no longer the most relied upon metric to value an intangible-heavy
business, it creates a very significant problem for private capital
markets as well as public.

To understand what a technology-based business may be worth,
and to fundamentally value the technology, consider:

> If you have two companies that do the exact same thing, but one
> is tech-enabled, and one is not, the tech-enabled business trades
> at a ten times EBITDA multiple. The non-tech-enabled business
> trades at a five times EBITDA multiple. All else being equal, one
> could claim that there are five EBITDA points of value in the
> technology asset versus the non-technology asset based on the
> anecdotal example I am providing you. This exists. You see this
> in public markets all the time. Tech stories carry higher multiples.
> Whether or not the market buys that tech story is a separate issue,
> which is our scope for this book.

All of this boils down to a basic problem, which is that cash is no longer the fundamental king or queen in capital markets. For centuries, capital markets have been driven by the age-old adage that cash is king and that a company's ability to produce cash is ultimately what gives it value. At the time of writing this book, we see this within many examples, even with the volatility in the markets today. Technology companies often produce no cash and have paid zero dividends. If you buy a technology stock, you will never get the money you paid for that stock back through any kind of cash disbursement or dividend; your only hope is to sell that stock later on through an appreciation of equity. So, cash is no longer king; it is the story that becomes king; it is the intangible assets that become king, which creates a very significant challenge for investors, business owners, capital markets executives, and regulators; it is a huge challenge.

And the premise of this book is that systems engineering tools can solve a very small part of this challenge, and they can enable engineers in particular. People who have expertise in technical fields understand how to measure and articulate the value of these non-EBITDA-producing investments or these non-EBITDA-producing assets, which are intangible to the market, whether that market may be customers, investors, or someone else.

It is important right now to establish a caveat or a model and to say that this example of how intangible asset-heavy businesses have higher EBITDA multiples because they generally have lower EBITDA is a correct example, and this is how capital markets work. The problem with establishing a paradigm such as this is that it can be twisted or manipulated.

It is really important for the systems engineer, the business owner, or the capital markets executive reading this book to remember that there is no substitute for a cash or revenue-producing business, and just because a firm may have all the right tools, all the right intangible assets, all the right software code, does not mean that firm should command a premium value in the capital market ecosystem.

The market ultimately tells businesses what they are worth based on who is willing to buy those businesses' shares on a primary or secondary basis and at what valuation. There is no magic bullet for building and running a business. Intangible assets are not an excuse for getting customers and developing revenue. Intangible assets are a method to get customers and revenue that, before 1995, barely existed and, at the time of the writing of this book in 2023, is extremely relevant and perhaps even the more common way to build a business.

The last perspective of how to value intangible assets comes back to something we talked about earlier with regard to goodwill. To value an intangible asset, you need a transaction. "You," being the investor. You need a transaction to review. There is a business. The business was purchased for a hundred dollars, and there is an analysis that says that 70 percent of that hundred dollars was in intangible assets, 90 percent of that 70 percent was in goodwill, and the remaining 10 percent of that 70 percent was in identifiable intangibles like copyrights, patents, trademarks, and so on. Goodwill is a plug.

Goodwill in and of itself does not need anything except to say that there is a premium – above what the purchase price allocation can determine – that indicates what the asset is worth. Goodwill can become useful if one does a highly statistical analysis and a data analysis to determine how much goodwill exists and how goodwill differs across different industry comparatives.

We will get into this in due course. We will talk about health insurance comparables, consumer product comparables, and ad-tech comparables.

The fundamental premise here for a systems engineer, a business owner, or an investor is that valuing intangible assets can be done through one of the methods that I mentioned, but also that it is fundamentally a very tricky method. Valuing, in general, even for tangible-asset-heavy businesses, is more art than science, and that adage holds even truer for an intangible-heavy business.

Chapter 7
Separability, Measurability, and Predictability

We have already discussed the absence of a single magic bullet for using intangibles to create an artificial value of a business; however, systems engineering enumerates a fairly clear path toward that valuation using three specific factors plus three specific tools. These factors and tools can allow a business owner or investor to identify and articulate the intangible assets that already exist.

The three factors I am speaking of here are:

- Separability;
- Measurability; and
- Predictability.

Separability is often the most difficult of these three to understand. A separable asset is not dependent on another asset for its value. It is distinct from other assets. For example, you may say that a person is good-looking, a somewhat subjective assessment, but that is simply based on the geometry of their face. Perhaps this person would be able to attract compensation for that attractive face as a movie actor because that face is attached to a body and a voice;

however, at least at present, a disembodied face, separate from body and head, cannot be valued as an asset. Separate a face from its head, and the core asset ceases to exist.

The phrase "I'm loving it!" is made up of three simple words, none of which are fundamentally owned by McDonald's, but when used together, these three words evoke a deep association with the brand. They are, nonetheless, separable from the brand and the company itself, valuable as a trademark or copyright.

Measurability is often overlooked or rushed through by business owners and investors. The measurability of an intangible asset has to be consistent, and it has to be consistent across an industry. The software industry, for example, typically assigns value to simple metrics like daily active or monthly active users, churn (a ratio of how many users terminate during a period of a contractual arrangement, compared to how many users you started that period with), and net revenue retention, which is an inversion of churn.

The number of patents a company holds could also be an effective measure of value. Imagine you have two companies that are competing, and they both have similar revenue and EBITDA and a very similar customer base – but one has a thousand patents, and the other one has two. A company with a thousand patents is going to be more competitive in protecting its products and operations in court.

These metrics are very popular intangible asset metrics in the industry, largely because they can be identified in just about any software company. If, however, a company develops a unique way to measure its intangible assets that no other company shares, you are pretty safe assuming that the measuring tool is invalid.

Predictability means the value of your intangible asset should be consistent over time, and the method of measuring that value should also be consistent. If there is a unique intangible asset that a company measures one year, and then does not measure that asset the following year, either that asset was sold or otherwise disposed of, or it was never a valid asset in the first place.

Given these three factors, we have three tools in systems engineering that are extremely useful. Those tools are:

- Requirements-tracing matrices;
- Failure mode and effect analysis (FMEA); and
- Interface analysis.

A requirements-tracing matrix is a familiar tool to systems engineers because it involves taking a technology or engineering requirement that has been drafted by an engineering team and then mapping it to other sub-requirements. The top level of the matrix is called the originating requirement. The originating requirement can be something simple, like "the remote-controlled car will be powered by direct access to the electronic grid." A derivative requirement could be that the power grid supply must be 12 volt.

Very often, an investor or capital market asks for data that are not easily measured. There is no easy button for finding out daily active users on a specific software platform, for example. The originating requirement is that your measuring system must keep track of a rolling average of the number of daily active users on an ongoing basis. It may sound simple, but there would be numerous derivative requirements if you aim to present a useful number. Daily active users have to be defined; they then have to be accurately measured over time. If you have a user that logs in for half a second and then immediately logs out, does that count as an active user?

The business owner has to address dozens of similar derivative requirements for each meaningful measure of value within their technology. It is important to be able to understand some of the complexities that underlie even the simplest metrics because the number itself may not mean much if you do not understand how it was derived.

Failure mode and effect analysis (FMEA) was introduced by the US Armed Forces in 1949 and popularized within the manufacturing industry by the mid-1970s. FMEA diagrams define the risk factors that are associated with a system – the ways that the system can fail. We are not talking about a universal failure, along the lines of a complete regional power grid failure, for example, because that is not a useful risk factor to attempt to manage.

An FMEA diagram for the metric for daily active users would describe all possible failure points that affect that measurement, for example, a user who tries to log in but is unable to. If your key performance indicators and your intangible assets are articulated with sufficient specificity and separability, they will all reveal failure modes. Understanding those failure modes will help you identify the way they affect the valuation of your intangible assets. You can then rank their frequency and severity of impact and address the worst offenders.

Interface tracking (also referred to as interface analysis) describes and manages the points of interaction between your system and other systems that provide data or functionality that your system relies on. Software engineers tend to understand the value of interfaces, and their literacy within the use and context of interfaces is very strong; however, all other professional profiles within a firm often seem ill-prepared when it comes to understanding the importance of carefully managing system interfaces.

An interface is something that would be described in your requirements-tracing matrices and also identified as a potential failure mode in your FMEA. Tracking interfaces provides perspective on what your system is capable of. It also helps you better understand what other systems you are dependent on and identify additional functions you could outsource. Quick, obvious note here: a business should never outsource its core competencies!

Interface tracking within the context of systems engineering specifically involves making sure everyone is following the same basic rubric for how they are measuring and defining the success of the system. The more stakeholders involved, the more important it becomes that interface tracking is carried out with discipline and consistency across all stakeholders.

All three of these tools – the requirements-tracing matrix, FMEA diagrams, and interface tracking – have some things in common: each one involves diagramming, analyzing, and understanding relationships.

A requirements-tracing matrix describes the relationship between a requirement itself and the many sub-requirements from which it is derived.

FMEA diagrams drill down into the relationship between the successful state of the system and all possible points of failure within that system.

Interface tracking describes the relationship between the requirements in your system, the data and functionality that are essential to it, and other systems that yours is either influencing or influenced by.

I wanted to develop this idea about relationships in part to distinguish the challenge of evaluating a business from, say, the challenge of solving a complex mathematical equation. In mathematics, you can change one variable in a seven-by-fifteen matrix and anticipate a clear, well-defined result. You cannot do that in a business.

For example, looking at your business through the lens of the requirements-tracing matrix, you will see pretty quickly that not only is each requirement dependent on several other variables but that there are potential failure modes within each of those variables. Solving one failure mode in a long chain of interdependent sub-relationships does not automatically boost the valuation of your business. Your solution may also highlight a new failure mode that you would not have otherwise discovered and may also stress a related interface in ways you could never have anticipated.

The complexity of defining all of this becomes extreme fairly quickly because each potential iteration becomes a long chain, a linear program of sorts. If you try to imagine it as a mathematical equation described by that same seven by fifteen matrix, no decision is a single, isolated variable. It is a variable among perhaps ten or fifteen other dynamic variables that are iterating and changing constantly. Many may even change independently while you are working on changing the initial variable because of the chaos factor of working with other human beings.

Before spending lots of money addressing the problems they assume are the most important, executives should implement a few of these systems engineering tools themselves, just to start to think in a more orderly manner about how all of these parts of their business come together to influence each other.

It may be helpful here to pause for a moment and think about what a bird's eye view of these elements might look like for your own business – or in a business you are evaluating – to consider how all these elements interact. You will start to be able to identify various intangibles, each separate, measurable, and predictable. You can consider how you might evaluate one of them using the three tools I have mentioned, creating a visual construct of the relationships between different intangible assets and the impacts each might have on the others and the whole system. The daily active user metric might be a good place to start.

As you begin to see the influence of various metrics, you have to figure out how to bring all of them together. To do this in an orderly, perfect picture is impossible. It will never work perfectly, but the benefit of systems engineering and the simplicity of some of the systems engineering tools we are discussing here is that they can bring you a few steps closer without trying anything particularly expensive or difficult. The exercise may also help you begin to identify some of the weak links in the chain so you can address them with a new, generative, considered approach.

Going back to our chief example, the number of daily active users is, at the very least, an important intangible asset. Daily active users are a metric that is separable and measurable, once you define exactly what constitutes that metric for your business. This metric is also relatively predictable. You are not going to have a thousand daily active users today and zero tomorrow. Tomorrow you will probably have a number similar to today's number.

Deciding what constitutes a daily active user is a function of your requirements-tracing matrix. If you are not sure where to start, you might consider a daily active user as anyone having at least thirty seconds of logged-on time within a twenty-four-hour period, with a derived requirement that a failed log-in attempt should probably also be counted among daily active users.

If this measure of failed log-in attempts is significant, you would want to execute a failure mode and effect analysis to understand the reasons your users are experiencing trouble logging in. Is it typically a simple human error like forgetting a password, or a security malfunction, followed by a failure of your password reset protocol? Was the user's account hacked? Was their data made vulnerable?

Once you identify the failure modes, you would use interface tracking to understand just where responsibility lies. Is there something amiss with your internal interface? Is the log-in process design not sufficiently intuitive? Or is there an external interface that is malfunctioning, like a reCAPTCHA tool that is too difficult for some users to navigate?

Following this process allows a manager, executive, or even an investor – without trying to boil the ocean and create seventeen different projects and without spending loads of cash – to break down complex challenges and understand what it might take to address them. You can see how even this simple example turns fairly complex once you begin to look at it systematically. Most problems in your business will be much more so, but you can be confident that you have the tools to get you moving down the right path, even for the most complex issues.

Now, the executive or the investor does not need to be the person to go to fix the username and password matching protocols or to rebuild the database after it goes down, but it does behoove them to at least think clearly about how these kinds of challenges can be effectively broken down and attended to. Each struggle is also an opportunity to create new strengths, impacting business outcomes and, ultimately, valuation.

Chapter 8
Why Systems Engineering Is Relevant for Corporate Valuations and Intangible Assets

It is important to understand the role that systems engineering plays in most engineering disciplines. Systems engineering is fundamentally a tool for integrating and connecting these disciplines. It is common in aerospace, and it is used to help mechanical engineers talk to electrical engineers and enable them to work in a seamless framework with physical engineers.

The systems engineering skill set is about bringing different languages together in a common framework. Since 1995, when the growth of intangible assets became prominent, technology has played an increasingly important role in the lives of business owners.

This book's subtitle could have been more related to technology rather than intangible assets. The reason I chose intangible assets as the subject matter is that intangible assets are measurable, and they are something you can apply measurability to, something you can apply systems engineering to, and something that you can help diagram within the techniques and the principles that I am communicating in this book.

Speaking about intangible assets helps you focus specifically on what you are talking about, whereas technology is itself a very broad term and could encompass anything from patents and copyrights all the way to software code and technical know-how and even all the way to the good luck that technology can include.

Systems engineering brings together these basic principles that, in some ways, are almost MBA-like. They are operational tools that we will get into in the course of this book, such as requirement traceability and failure mode and effect analysis. They are not fundamentally different from what one may learn in an MBA program.

The difference between systems engineering and your classical business curriculum is that systems engineering is fundamentally about measuring physical or virtual systems – it is about measuring outputs. In the MBA curriculum today – which, in my premise has failed to adapt to the modern workforce needs – it becomes very high level. The MBA skill set has largely been replaced by Google, and the readily accessible information that exists in technology is such that the tools that a person picks up from an MBA program for intangible assets are pretty useless.

On the tangible asset side, the MBA programs are still very competitive. Financial statement analysis, securities understanding, and financial principles like accounting are all designed to measure tangible assets, so MBA programs are, by and large, the best option in the educational market to investigate these ideas. But there is no sophisticated curriculum for intangible assets or technology.

My premise is that systems engineering presents tools and techniques that business owners, investors, and anyone in the global ecosystem of business can use to start thinking about intangible assets as they relate to valuations in a more concrete manner.

If you went to school before 1995 – and in many cases, before 2010 – and you did not study engineering, you missed out on understanding how to really measure, identify, and improve technology through the same kinds of abstract tools that are made available inside of business programs across the United States and the world. That is fundamentally why the systems engineering pedagogy is useful for intangibles and within the use case of corporate valuations.

My premise is that the tools we present to the business owner and the reader inside this part will be immensely valuable – not necessarily for putting into PowerPoint slides or Excel models that then present to investors, but for helping business owners or investors to break down what the technology is doing by defining the measurability of different outputs that they deem to be valuable. Finally, you will be able to overlay those outputs that you have deemed valuable into what the market ultimately pays for, which gets back to the crossover between intangible assets and corporate valuations.

Chapter 9

What Intangibles Matter Most for Corporate Valuation

Part I of this book is titled "Intangible Assets" and in Chapter 4 we discussed how to define intangible assets and exactly what they are. In Chapter 5, we dealt with how to value intangible assets and how internal and external factors influence that valuation. Chapter 5 also explored how to put these defined and valued intangible assets to work inside a business in terms of utilizing them to drive corporate valuation and also, of course, in business operations scenarios. Soon, we will arrive at Part II.

Part II will deal specifically with the tools that will give you the outcomes we are exploring regarding valuing intangible assets. But first, let us deep dive into valuation so the tools in Part II will be more relevant.

We will review what constitutes a company valuation, why they matter, and how they differ among different kinds of companies and in different scenarios. We will address items directly related to intangible assets, and we will identify some items that you must consider when developing valuations, including some that are not related to intangible assets.

So, what intangibles matter most when it comes to business valuation? Throughout the book, we have anecdotally reviewed many intangibles, such as the good-lookingness of an executive team, daily active users, trademarks, patents, copyrights, and other various identifiable and unidentifiable intangibles.

The intangibles that matter to a valuation are fundamentally the ones that the buyer will pay for, and the intangibles that move the needle most are the ones that the buyer will pay the most for. This may seem obvious or simple, but it is often overlooked when valuing a business.

Another fairly obvious factor that is often overlooked by investors is the reason for considering an acquisition in the first place. What type of acquisition transaction are we dealing with? There are two basic transactions that we will explore. You also have to consider the timing of the transaction.

Within the linear program of systems engineering, intangible assets, and corporate valuations, we have now created an additional set of variables that will need to find their place in the computational matrix.

If a business owner, investor, or researcher wants to accurately value a business, they must start by understanding the motives of the buyer. That means, of the ones I have listed, the most important frame of reference is the type of transaction that is directly related to motive.

For our purposes, we divide transaction types into two categories: primary and secondary. A primary transaction means cash flows directly to the balance sheet of a business in exchange for ownership shares in that business. In a secondary transaction, that cash is sent to a shareholder instead. You can see fairly quickly that the main difference between these transactions is one of the motives. In primary transactions, intangible assets are valued at a greater premium than in secondary transactions. We will talk about why in Part II.

Motive, incidentally, is also the best lens through which to discuss the differences between venture capital, private equity, and public equity. A quick note here: for the rest of the book, and whenever we talk about valuations, we are exclusively talking about equity. I have avoided talking about security instruments like convertible debt or derivatives because they simply serve to layer-on complexity. If you understand intangible assets, systems engineering, and how they affect corporate valuations and equity investments, you can extrapolate those learnings into whatever securities instruments you typically deal with.

The first thing we need to understand is that primary and secondary transactions influence the valuation of intangibles differently. The second factor that affects the valuation of intangible assets is the scenario of the transaction. The scenario typically refers to whether we are looking at a capital placement – like a minority capital raise that grows the asset – or are we considering an M&A sell-side or buy-side? There are other possible scenarios, but for the most part, we will focus on just these three.

The last element when deciding which intangibles matter most is related to the industry within which the business operates, or you could also describe it as the business use case for the transaction. We now have primary and secondary transactions, we have capital placement, M&A sell-side, M&A buy-side, and the business case for each of these different elements of the transaction (see Figure 9.1).

The business case of the primary transaction for the sell-side is that you want to communicate a valuation that is high because buyers in your industry want to acquire your intangibles, and you are generating cash to reinvest. If you are courting a cosmetics company, they want your customer relationships, and they want access to your formulations. If it is a plastics manufacturer, it may be investing to benefit from access to your supply chains and your

	Capital Placement	M&A Sell-Side	M&A Buy-Side
Primary	Motive: Strengthen balance sheet, re-invest	Motive: Unwind venture capital and reward investors, scale	Motive: Consolidate resources
Secondary	Motive: Divest and re-allocate, diversify	Motive: Retire	Motive: Consolidate Decision-Making

Figure 9.1

relationships with vendors and suppliers. Your ability to strengthen the balance sheet of your business, or to invest in research and development (R&D) or new technologies, will be dependent on how much cash you are able to raise. Money from a primary transaction is not "personal retirement boat money." It is asset-growth money that the buyers expect you to invest in such a way that their equity gains value.

These are fairly safe assumptions with regard to a primary transaction on the sell-side, whether we are talking about capital placement or M&A. These assumptions change when you start talking about a secondary transaction on the sell-side. No longer are we considering doubling down on the business. Now, we are talking about that boat money! You are gathering cash so you can retire, reallocate, or diversify your personal portfolio. Now your motive as a seller is completely different. This is your last chance to experience the value of your asset. After the sale, it will belong to someone else, who will become the exclusive shareholder of the equity you have released in the sale. In some cases, the recipient of that value becomes the business itself or a combination of many shareholders.

The industry within which the transaction falls creates custom business cases for each buyer, as I alluded to with the examples of the cosmetics company and the plastics manufacturer. Researching specific industries may help you understand which of your intangibles matter, when they matter, and to whom they matter most. Insiders and executives in various industries often treat intangible valuation within their industry as common knowledge, and we will talk more about that later.

Let us quickly look at a similar analysis of primary and secondary transactions on the buy-side. A buy-side primary transaction means an investor is buying shares of a company, and that capital is going directly into the company to sit on the balance sheet of the company for growth. In this case, the intangible assets that matter most to buyers are the ones they believe will create an outsized return on their investment over time. They assume there is already some set of intangible assets that exist inside this company and that their investment will augment these intangible assets at a rate disproportionately high compared to the amount of money that they are putting in and will then command a higher value in the secondary market later on. As you stack the considerations on top of one another, you get way beyond some kind of simple binomial equation. We are considering at least three variables, more likely four, sometimes even five.

Most people working through a valuation will ignore the difference between primary and secondary motivations. They are working on the assumption that all valuations are basically the same and typically gravitate toward the context of secondary transactions without understanding how motives can be skewed on a primary basis. When you are looking at a primary transaction on the buy-side – perhaps an investor is putting money into a company on a minority basis – that investor does not want the valuation to be too high. They see the value of intangible assets, but they primarily want to optimize the valuation they are coming in on compared to what they believe will be the valuation in the future.

On the sell-side of a primary transaction, the seller may have a range of motivations. They would certainly hope to use the investment to increase the future value of the business, and the higher the valuation at the time of the transaction, the more resources they will be able to leverage toward scaling per unit of surrendered equity. They would also hope to realize a valuation that is reasonably related to a secondary transaction they expect to benefit from later on; because, ultimately, they hope to increase their personal wealth as much as possible.

As you can see, some of the motives of the buyer and the seller in a primary transaction are more or less aligned, while some are opposed; however, in a secondary transaction, the motives of the buyer and seller are typically opposed. The seller in a secondary transaction wants to maximize their personal wealth, and the buyer in a secondary transaction wants to minimize their outflow of cash. There is always this tension between buyer and seller in a secondary transaction, especially when we are dealing with a merger or acquisition. M&As are almost always secondary transactions, although you will sometimes have a mix of primary and secondary transactions.

You do not have that diametric opposition on a primary basis. On a primary basis, you have a seller who wants to sell a security and communicate a value of an intangible asset. They will want to communicate a value that they feel is fair based on what they expect to achieve or what they have already built. The investor wants to communicate a value that is also fair, but that also allows a margin between the valuation that the investor is coming in on today and what the investor believes they can eventually get in the secondary market.

This margin of interpretation between the buy- and the sell-side is really what leads to these very large and often unreasonable valuations in a lot of primary transactions, particularly ones that we see in the venture capital (VC) market. It is also what leads to overly conservative valuations in other markets. The lack of transparent reporting and lack of understanding of intangible assets creates a big "I don't know" factor for parties on both sides of the transaction. And that "I don't know" leaves too much room for interpretation when the ultimate goal is to see sellers and buyers find a price point that is maximally mutually beneficial.

The result of the "I don't know" factor is that sometimes, you will notice that some companies only attract equity investment on a very conservative valuation, like those you often see on Shark Tank or Dragons Den. Shark Tank entrepreneurs often do not know what their business is worth, and you will seldom see one who seems to have a thorough knowledge of intangible assets. Other times – think Silicon Valley and VC – you will see a company sell at what seems like a ridiculously bloated valuation. These tend to be companies that do understand the value of their intangible assets and also understand how to inflate their perceived value.

At the end of the day, regardless of whether the primary transaction represents a conservative or aggressive valuation, both buyer and seller hope that the company's intangible assets will lead to tangible results that command a premium in the secondary market – a premium to what the seller sold at, and a premium to what the buyer bought at.

The three considerations we have discussed are: the type of transaction, either primary or secondary, then the scenario of the transaction, which we defined as minority capital placement, M&A buy-side, or M&A sell-side, and finally, the industry or the business case for the transaction. I will address the business case in more depth later. Although this information can be confusing and may feel like it is a lot of investment banking terminology, it is essential in order to understand what intangibles matter because they matter differently based on a combination of the variables.

One additional consideration that comes to bear is the general profile of the buyer, including the buyer's relationship to the industry and interest in the specific business case for the transaction. In the world of buyers and investors for investment banking, there are three major categories: private equity, venture capital, and strategic incorporation.

Private equity investors occasionally participate in primary transactions, but more often, they deal with a secondary transaction. They typically buy or invest in private companies based on established cash flow and EBITDA multiples. The intangible assets matter, but they matter in a binary way. The business under consideration either has them or does not. If it has them, great. Investors are more likely to buy. If it does not, they are less likely to buy; but the valuation itself is ultimately driven by EBITDA, not by intangible assets. Private equity is wired that way, and it will never change.

Venture capital, on the other hand, is all about intangible assets. Venture capitalists (VCs) are looking for very innovative companies they expect to grow at a high rate. They expect the business to spend lots of money to grow, but they see that growth is primarily driven by considerations that cannot be communicated on a financial statement or through cash flow analysis. When VCs see intangible assets they believe will drive growth, they place a high value on those intangibles.

Strategic incorporation refers to a business that wants control of new assets to enhance an existing business line. Strategics are industry-specific, and within that industry, they are also intangible-specific. Industries know which intangibles are most valuable to them, so they will value specific intangibles very highly while disregarding others that may be more valuable to a different industry. The prosthetics industry shares some needs with the health insurance industry. They both value access to a large user base within healthcare ecosystems, but a dependable supply chain for carbon fiber fabric and stainless steel is more important for one, while regulatory concerns may be more important for the other.

The intangibles that matter most change on a constant basis. In some industries or at some periods, they may even change on a daily basis. They can be influenced by customer behavior or the season's cotton harvest. They evolve. One of the examples we will look at a bit later is why health insurance companies value Medicaid assets more than Medicare assets or other health insurance relationships, specifically because of regulatory reforms that were sending more money to Medicaid programs, making Medicaid patients more valuable.

Internet speed and computing power used to be crucial price-setting intangibles. They do not vary across industries now the way they used to, and they do not affect valuations in 2023 the way they did in 1999. Retail presence was a must for most brands prior to the global lockdown that began in 2020.

With the entire world stuck at home and pivoting to mostly shopping online, in a matter of weeks, every investment in foot traffic became a liability. Intangibles change faster than tangibles, and in the economy that we in the West live in – which is driven by the Internet and highly dependent on the ability to develop new technologies – assets are increasingly intangible.

* * *

It is important to have a call out here about how valuations impact corporate expansion. Corporate expansion is an important element among the principles of systems engineering that business owners, investors, and researchers can use, but a corporate expansion use case is specifically about revenue and communicating value to customers.

Intangibles and valuations are significantly impacted by corporate expansion. Corporate expansion brings more tangible financial results. If historically, your business only serves the US market, and then you expand to Canada and the UAE, that increased market presence should increase your overall net income or your EBITDA, which should increase your overall tangible valuation.

In addition to the tangible results, the intangible results include more customer data, more complex customer data, more comprehensive customer data, and a proven use case for a larger market, all of which further increase potential valuation. Perhaps you could then expand into Saudi Arabia and China while also targeting expansion into developing markets in Indonesia, Egypt, and India. Now you have a product that is proving itself in the larger global market, and your product will be more valuable than a product with the same key performance indicator (KPI) generated from a smaller market.

So, intangibles and valuations are impacted by corporate expansion in part because they are the outcomes of corporate expansion. It is often customer interest in intangible assets that drives corporate expansion, which again expands your intangible assets, which, assuming all of this, is done with sufficient profit and drives higher valuations. The risk of corporate expansion is that low performance of that expansion drives lower valuations. Just as a similar KPI in a wider market can drive an outsized increase in valuation, a lower KPI in a wider market can drive an outsized decrease in valuation.

There is both an art and a science to evaluating and marketing intangible assets. Selling them is not like trying to sell beer to frat boys at a packed bar. They are all thirsty, the price is the same for every buyer, and the main thing they want is more. When you are selling a business, every buyer is different, and the intangibles must eventually, predictably, drive tangible results beyond just getting a buzz. The valuation tools you use to assess the business will be different for each buyer.

Occasionally you will hear about just the right kind of intangibles finding just the right kind of buyer, and a company goes from a million or two in revenue to a billion-dollar valuation overnight. A "unicorn"; you hear about it precisely because it is so rare. Do not build your future around that goal. Instead, if you are a business owner, the goal is to communicate clearly about your intangible assets, so you will have the best possible opportunity to move the needle in the right direction.

Chapter 10
What Intangibles Matter Most for Corporate Expansion

Until now, we have focused on the impact of intangible assets on corporate valuation and on how intangible assets can be further explained or understood using systems engineering tools. We have approached primarily from the perspective of an investor and what results they look for in terms of a firm's tangible and intangible assets because corporate valuation is fundamentally a negotiation with purchasers of ownership stock or business debt.

There is another important negotiation that is constantly happening, and that is the negotiation between a company and its customers. That is why no work on corporate valuation would be complete without also taking a deeper look at corporate expansion.

The motivations of an investor who purchases the stock or debt of a business are different from the motivations of the customer who uses its products. Corporate expansion is only possible as customer demand for those products increases. I cannot tell you how many times, in my experience of working with business owners across the world, they fail to understand the nuances that differentiate the customer from the investor. CEOs and business owners who have become very successful and have scaled businesses, however, tend to understand very well what customers want and how to communicate value to them.

Talk to any CEO of a successful company, and you will find they are invariably extremely eloquent in all things related to their customers, what their customers have, what their customers want today, what their customers may want tomorrow, and they also tend to understand pretty well how their customers are paying for each of those things. If they have had success in one market, the very next consideration is what might be required to increase customer usage of their products or to capture new customers in a new market altogether, either domestically or internationally.

The first thing you might do when entering a new market is to conduct some version of a demand survey to try to understand what customers in that market are saying they want. Sophisticated leaders understand, however, that there is often a marked difference between what potential customers say they want, what they are willing to pay for, and the fundamental costs and unit economics of what you are trying to sell them.

So, when do intangibles matter in corporate expansion? Remember, we discussed that the influence of intangibles on valuation is industry-dependent. Similarly, the influence of intangibles on corporate expansion is dependent both on the industry that a business is a part of and on the customer base that they are considering. Let us look, for example, at a business in the fashion domain.

Celebrity endorsements largely influence customers in fashion. If a specific celebrity wears your brand, the fans of that celebrity are very likely to be attracted to your brand. We had a client in the footwear industry who sold luxury African products. They had wonderful celebrity endorsers wearing their products. But most of the potential customers in one market we looked at had little to no awareness of that specific celebrity. That endorsement is not an intangible asset sufficient to justify corporate expansion in that specific market because it is not an intangible asset that the customer is willing to pay for.

There is also a regional component. In my opinion, the trading blocs experiencing the most growth today are the Middle East, North Africa, South and Southeast Asia, and Latin America. Gross domestic product (GDP) dynamics, birth rates, and other economic drivers make these regions compelling during the coming decade. Whether your firm is headquartered in the United States or elsewhere, you have several ways to approach corporate expansion in these regions, including joint ventures, strategic partnerships, acquisitions of companies, and good old-fashioned sales or distribution agreements. But each of these regions will pose specific challenges to expansion.

Let us look for a minute at the example of the Middle East/North Africa (sometimes referred to as MENA). When firms consider a global expansion to MENA, they need to understand how diverse this region is. MENA broadly refers to over five hundred and fifty million people that speak over seventeen dialects of Arabic – who are spread across a very large region with several large, rapidly growing global cities – with a population and influence concentration in Egypt and Saudi Arabia.

Economic strength in these countries is concentrated around the oil resources of the Gulf Cooperation Council (GCC), namely Oman, Bahrain, Kuwait, Qatar, Saudi Arabia, and the UAE. Historically, over half of the GDP of GCC countries has come from oil production. The uncertainty around oil demand during the recent pandemic forced these countries to redouble efforts toward economic diversification, as seen in Saudi Arabia's Vision 2030 and other initiatives by the UAE, Qatar, Saudi, Oman, and Kuwait. The big struggle is that any significant investment by these countries in economic diversification still has to be funded by oil revenues. We will explore these regions in more depth in Part III.

Chapter 11
Buy-Side Intangibles

Industry Examples of Buy-Side Intangible Assets and When Intangible Assets Matter to Buyers

Until now, we have examined the utility that intangible assets play in corporate valuations on the sell-side. We have examined the sell-side, primary transactions, and secondary transactions. Now we are going to examine buy-side secondary transactions.

We will discuss when intangible assets matter, particularly to private equity buyers, and how private equity buyers use intangible assets to maximize corporate value.

To address these questions, we reviewed six thousand purchase price allocations from publicly traded companies to determine what intangible assets matter most by industry. We conducted this analysis of M&A activity between 2010 and 2018, particularly among the buyers in five verticals. The verticals were financial services, healthcare, energy, information technology, and branded consumer products. We looked at all of their purchase price allocations, and we broke down the amount attributed to intangible assets by industry.

The research indicated that customer relationships held the greatest percentage of intangible value for healthcare, followed by financial services, and ironically, the lowest amount of customer relationship intangible value existed in branded consumer products and information technology.

Information technology and branded consumer products had the highest allocations attributed to developed technology – which is the technology that is developed for internal use.

When looking at these six thousand purchase price allocations, we considered how much of the identifiable intangibles existed inside of the use case.

What we were able to determine was that identifying the most intangibles created the most corporate value.

Publicly traded companies have to test their goodwill for impairment in their financial reporting, and goodwill is impaired when the acquired intangible asset loses its value at a rate faster than what was initially perceived.

Private equity companies do not have to test goodwill for impairment, and therefore, they are free from some of that burden.

Overall, the clear indication for these industries was that intangible assets matter in amounts of billions of dollars, and the kinds of intangibles that mattered by industry are surprising, as you can see in Figures 11.1 and 11.2.

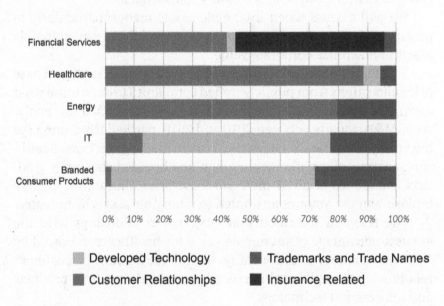

Figure 11.1

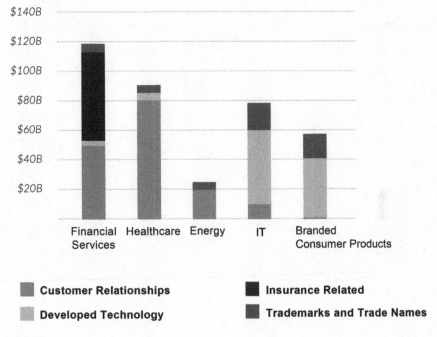

Figure 11.2

Chapter 12
PE Ratios

It is important to understand how intangible assets are valued by public market investors – one of the best metrics available for understanding that valuation is a price-to-earnings (PE) ratio on publicly traded equities.

Simply defined, a PE ratio is the price that a stock trades at divided by the number of earnings per share that the stock generates based on historical trends.

One can intuitively understand that when a PE ratio deal is very high – meaning you have a price that is approaching infinity, whereas the earnings stay relatively low – then there is something that the buyer of the share values that is not reflected solely in the earnings; particularly when juxtaposed to a PE ratio that is very low, meaning you have a price of a stock approaching zero and the earnings staying relatively constant.

One can visualize these two examples and extremes via Figure 12.1. If you have price over earnings then you have a stock of intangible value, with a price of a billion dollars for one share and earnings of one dollar, and then you have the price for a low ratio (a tangible heavy stock), with a price of one dollar and earnings of one dollar, then there is a scale of valuation attributed to intangible assets.

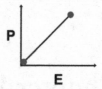

Figure 12.1

This is particularly true when you start analyzing sectors. If one investigates the PE ratio of banks according to data published by the Stern Business School in January 2003, banks have PE ratios of around seven, whereas advertising can have PE ratios of over a thousand depending on what the sample set is.

The most stable (or boring) kind of example is manufacturing. When one investigates, one sees that manufacturing has a PE ratio of eleven to fourteen. For precious metals, retail, and so on, the ratios are often below twenty but in software for entertainment and Internet applications, one can see that the PE ratios are fifty and sometimes even four hundred. This is based on an industry aggregation metric.

Tobacco may be one of the best examples of a stabilized ratio. Tobacco is fundamentally an addictive product. It is regulated, which means it has a high barrier to entry. Tobacco companies generally pay very high amounts of returns in dividends, as we see the publicly traded tobacco companies, and tobacco, as an industry, has a PE ratio of twelve.

Anything below tobacco would be seen as a very low intangible asset value, and anything above – and the more above it is – becomes a higher value of intangible assets. The bottom line is that when a PE ratio goes up, it means that investors or buyers of the shares are paying for something that is not the earnings, and by a process of elimination, there are really only two things an investor can pay for: One is they can pay for intangible assets, and the other is they can pay for growth, which is quite candidly a function of an intangible asset since intangible assets often refer to anything that is not earnings or that is not tangible.

Therefore, PE ratios are one of the most effective tools that we have in the public equity markets to understand how investors are responding to intangible assets on a global aggregated basis. PE ratios are not going to indicate if an investor is valuing software more than patents or patents more than copyrights or copyrights more than industrial growth. None of that is going to be clear in PE ratios. PE ratios simply point to the aggregate volume of intangible assets that investors put onto a business asset when compared with pure tangible assets.

Chapter 13
The Business Moat as an Intangible Asset

A business moat is another example of an intangible asset. A moat, in the context of medieval fortresses, refers to a protective barrier built around a castle to keep enemies at bay. In business, a moat is a metaphor for the unique competitive advantage that a company possesses, which creates a barrier for its competitors to imitate or surpass. These strategic moats give companies the ability to maintain their market share, command pricing power, and ultimately enjoy long-term profitability and success. This chapter delves into the concept of a moat in business, its importance, and the various types of moats that companies can build.

The importance of a business moat:

Protection against Competition: A moat in business safeguards a company from the threat of new entrants and existing rivals. By providing a unique value proposition or a product/service that is difficult to replicate, companies can maintain their competitive edge and prevent others from eroding their market share.

Enhanced Profitability: A well-constructed business moat allows companies to charge premium prices and enjoy better profit margins than their competitors. This increased profitability provides the financial resources to invest in innovation, growth, and other strategic initiatives, which further strengthen the company's moat.

Sustainable Growth: In the absence of a strong moat, rapid growth often comes at the cost of profitability, as companies need to slash prices or increase marketing spend to capture market share. A robust moat enables companies to grow sustainably without compromising on profitability or market share.

Investor Confidence: A business with a durable moat is more likely to attract and retain investors, as the company's competitive advantage provides them with a sense of security regarding the sustainability of its financial performance. This confidence translates to a higher valuation and better long-term returns for shareholders.

There are many types of business moats:

Intangible Assets: Intangible assets, such as patents, trademarks, and licenses, can provide a strong moat for companies by protecting their proprietary knowledge and preventing competitors from easily replicating their products or services. Intellectual property rights create a barrier to entry for new entrants, allowing the incumbent firm to sustain its competitive advantage and charge premium prices.

Example: Pharmaceutical companies that hold patents for their drugs enjoy a monopoly on the production and sale of those drugs for a certain period. This moat allows them to charge high prices and earn substantial profits until the patent expires.

Network Effects: The network effect occurs when the value of a product or service increases as more users join the network. Companies with network effects benefit from increased customer loyalty and a self-reinforcing growth cycle, making it difficult for competitors to catch up.

Example: Social media platforms such as Facebook and Twitter enjoy strong network effects, as the value of the platform increases with each additional user. This moat makes it difficult for new social media platforms to attract users and compete with established networks.

Economies of Scale: Economies of scale arise when companies can produce goods or services at a lower cost per unit as production volume increases. This cost advantage allows companies to price their products more competitively or enjoy higher profit margins.

Example: Walmart, with its massive distribution network and large volume purchases, enjoys economies of scale that allow it to offer products at lower prices than its competitors, thereby creating a moat that is difficult to replicate.

Cost Advantages: Cost advantages, independent of scale, can also act as a strong moat for companies. These advantages can be achieved through exclusive access to cheaper raw materials, vertical integration, or superior operational efficiency.

Example: In the airline industry, Southwest Airlines has built a moat by maintaining a simple fleet structure, efficient operations, and a low-cost business model. This advantage allows the company to offer competitive fares and remains relatively insulated.

Chapter 14

How Investors Think About Intangibles

I want to take a moment to share my insight on how investors think about intangible assets and what is needed in regard to intangible assets and their utility in accordance with an investor's point of view. To do this, we have to understand how different investors are profiled.

Investors can broadly be broken into two major categories and then into subcategories. The major categories are equity and debt. Equity investors (E) receive a return through three simple scenarios: An acquisition of their shares, dividend payments based on their equity ownership, or an Initial Public Offering (IPO), which is essentially an acquisition but with differences.

Debt investors (D) receive their money through interest payments. There are different mixtures or hybrids of these two groups, but for the most part, these are the major categories.

Equity investors are relying on intangible assets to create more returns in the future. Debt investors care much less about intangible assets. Debt investors care about the cash flow of a business and the ability of the business to pay the interest and return the capital via payments or a liquidity event. Within this multivariable equation, you have two of the variables here, E and D. There is another program in the matrix that matters, and we have talked about this in other examples for investment banking scenarios, but they become very relevant from the perspective of investors and those two other variables are E primary and secondary and D primary and secondary.

It is important to remember that a primary investor is someone who is putting money directly onto the balance sheet, meaning all equity investors exchange cash for some kind of equity security. Where the cash goes is what drives the determination of primary versus secondary. In a primary acquisition of equity, the cash goes directly to the issuer: the business. It sits on the balance sheet of the company and is used to develop or expand intangible assets for growth. For example, an IPO is a primary transaction because when investors buy those shares on the public markets, that capital then goes to the balance sheet of the company that issued the shares.

M&A is an example of an equity acquisition that is secondary because, in almost all cases, if a company sells its equity to a buyer in a controlled transaction, then the shareholders of the company receive the cash, and the company itself only maintains working capital. Debt can be performed on a primary or secondary basis too. Debt investors who invest on a primary basis are your typical lenders, such as banks, mezzanine funds, and commercial lenders. They wire money to the balance sheet of a business, and in exchange for that wiring of money, the debt investor receives a piece of paper that represents the debt security they own, which has a certain repayment plan, interest rate, and so on. You can trade debt on the secondary market but it is very rare for this to happen in private markets, this almost always happens in public markets. A good example is companies that own bonds issued by a large Fortune 500 company may sell those bonds to another investor if they feel like there is some kind of benefit to doing so.

These two different dimensions and these four different variables must be understood in order to evaluate the intangible assets that matter to an investor. The bottom line is that the debt investor is generally not going to be interested in intangible assets. They might be interested in intangible assets if they do some kind of hybrid structure or if the debt investor is a venture debt investor, but in those kinds of cases, the debt investor is playing a double-sided role. They are buying both equity and debt, and the amount of intangibles they care about really makes them an equity investor, whereas the debt decisions are almost always driven by the income statement, the cash flow, and the balance sheet.

For equity investors, intangibles matter a lot. On a primary basis, the intangibles are extremely important. Primary investors are almost always investing in mostly intangible assets. This is because they are assuming that, with their one million, ten million, one hundred million, one billion dollar investment that goes directly to the balance sheet on a primary basis for the issuer, the company will be able to turn that one, ten, one hundred million, or one billion dollars into more to give them a positive return on their investment by growing the business, and that has to happen through some kind of an intangible process. It has to. Otherwise, it is just a currency exchange.

On a secondary basis, there is a different dynamic in play. In a secondary transaction, investors care about intangible assets, and they do want them, but they are not as willing to overpay for them, or they do not value them as highly in secondary transactions because the money that they are spending will not, in turn, be used for intangible assets. In a primary transaction, the money is used to grow the business. It is used to develop intangibles. In a secondary transaction, the money is given to the shareholders to go buy a boat, so the money is not staying with the company to grow the investments.

There are an extremely high number of variations on how these intangible assets can impact investor thinking, but it is important at this moment to create and provide this summary of the intangible assets that matter to investors and how they matter, and how the way that they matter changes based on the transaction being considered.

Chapter 15

Intangible Assets in Branded Consumer Products

The last example that I want to talk about with regard to the "intangible assets that matter" in certain industries is the consumer product businesses. There was a company whose name I need to keep private but let us call them XA Labs. XA hired Jahani and Associates, who advised them on how to build and sell a company within five years. XA was the subsidiary of a parent company that generated fifty million dollars of revenue in nutraceuticals, and they wanted us to determine which assets were most applicable when selling the business in five years.

J&A discovered that in a branded consumer product business, acquisitions are most like ad-tech acquisitions because they allow companies to reach a specific customer.

We looked at three hundred M&A deals for branded consumer product companies. These companies included FDC vitamins, Life-Gen, Agilent, and Anacor. We looked at the acquisitions of the companies that were purchased by the companies from this list, and we picked twenty-five as part of a subsidiary concept to determine what the valuation was and how the valuation was allocated.

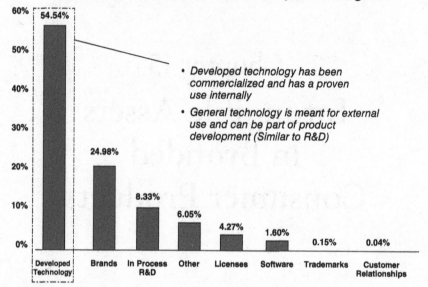

Percentage of Purchase Price for Acquired Intangibles

- *Developed technology has been commercialized and has a proven use internally*
- *General technology is meant for external use and can be part of product development (Similar to R&D)*

Figure 15.1

We broke down the purchase price allocation, as you can see in Figure 15.1, and we determined that developed technology was allocated for 54 percent of the entire purchase price spent on these twenty-five acquisitions. This represented hundreds of millions of dollars.

Customer relationship, ironically, was very low, as was trademark, which for a branded consumer product like a supplement, a coffee brand, or a shoe, to have trademarks and customer relationships so low was a surprise.

What we found consistently was that over half of the purchase price allocation was for developed technology, indicating that perhaps it was easier to acquire new customers than it was to develop new tech successfully. There are always fascinating secrets hidden in data that you will not find unless you are not looking for it. What I mean is, we are continually surprised by what we find, and it is the process of analysis that reveals counterintuitive facts.

Now, the developed technology associated with the 10-k filings that we were reviewing was internal technology. When a company develops technology to sell or to license, that is a different model from what we are discussing here. Developed technology for the purchase price allocation in these cases is specific to technology that is used internally, and it is not licensed.

This is really interesting because when you think about a shoe company or a water bottle company, there are many alternatives in this branded consumer products vertical, but despite the numerous alternatives, we see certain companies skyrocket to success. The reason for that is – based on the purchase price allocations – these companies own technology that allows them to reach their customers more efficiently than their competitors.

It is not about the trademark, it is not about the logo, and, counterintuitively, it is not about the customer relationship. You do not have a relationship with your ChapStick, your Hydro Flask, or your Nalgene water bottle. Nalgene just knows how to find you when you want a water bottle.

This could be through retail or online. What is so important about this example is that, again, for the intangible assets that matter most, it is not always obvious, and it is not always intuitive what companies really pay for. How they communicate to investors what they are paying for is a very profound way to discover what they really care about.

So, in summary, the most valuable intangible asset for branded consumer products companies is the technology that is used for marketing purposes to reach their consumer.

Chapter 16
Intangible Assets
in Healthcare

We have already covered the supporting material regarding intangible assets in ad-tech.

Now let us review the types of intangible assets that exist within health insurance industry companies. I conducted a study to understand the intangible assets that matter the most in the health insurance industry. The reason I chose to analyze the health insurance industry is that health insurance is complex; it is highly intangible driven, the companies are very cash rich, and it is a lucrative industry. Also, health insurance is somewhat unfamiliar to businesspeople outside the vertical. I reviewed all of the M&A acquisitions from 2010 to 2017 for the following buyers: Aetna, Anthem, Centene, Cigna, Humana, Magellan, Molina, UnitedHealthcare, and WellCare.

After researching these acquisitions and analyzing the purchase price allocations as reported in publicly available Securities and Exchange Commission (SEC) filings, I was able to calculate that 90 percent of the M&A value for all of the acquisitions by these buyers was assigned to intangible assets. Out of this 90 percent, customer-related intangible assets accounted for 65 percent.

The research pertains to specific intangible assets that are related to financial performance. What is important to understand about health insurance is that not all customer relationships in health insurance are the same. In the relationship between a health insurance company and a customer, what makes it lucrative is ultimately the amount of money that is available for a premium compared to the amount of money that is paid for that premium.

What this research showed was that in these acquisitions, remembering that 65 percent of that 90 percent was spent on customer relationships – the frequency of acquisitions was much higher for anything related to Medicaid.

So, let us take one step back. In the United States, there are three major components of health insurance. The first is the private market. In the private market, you can self-pay if you are a large company, or you can buy commercial plans available from the companies that I just mentioned.

Medicare is a government-subsidized health insurance that is available for people of a certain age. The third category is something called Medicaid, and Medicaid is available for people that are lower income, people that are within a certain percentage of the federal poverty level (FPL).

When looking at these acquisitions, I was able to determine that the number of acquisitions was very high for Medicaid, and this came down to regulatory intervention. You can see in Figure 16.1 that in 2009, Obamacare (The ACA or Affordable Care Act) was

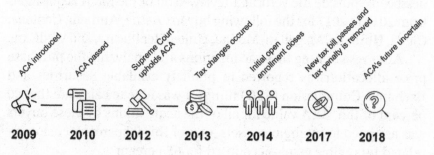

Figure 16.1

introduced, and it became formal law in 2012 after being challenged by the Supreme Court. One of the things that Obamacare did was increase the federal poverty limits for Medicaid. When it increased the federal poverty limits to be eligible for these Medicaid programs, it effectively increased the market size for these programs, therefore, increasing the number of customers that suddenly were eligible for a Medicaid health insurance product.

Medicaid is itself a relatively low-cost product. It is difficult to make a Medicaid plan extremely lucrative for the health insurance company because the government is very specific with how they want you to spend money. So just like any product that has a lower margin, the way to make it more profitable is to consolidate. This includes consolidation of IT systems, human capital, regulatory and reporting requirements, and so on, and that is exactly what happened.

In this sample set from 2010 to 2017 of health insurance giants, 90 percent of the money was spent on intangible assets. Of that 90 percent, over half of it was in customer relationships, and for all of the acquisitions, the most frequent acquisitions were made with regard to the Medicaid plans.

These same Medicaid plans – Centene and Molina in this example – also experienced the greatest revenue growth (Figure 16.2). Centene and Molina made twenty Medicaid acquisitions, and they

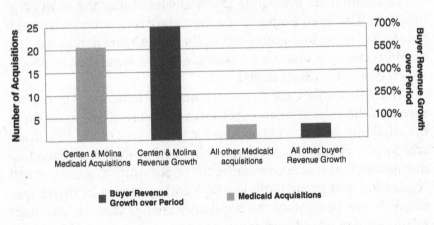

Figure 16.2

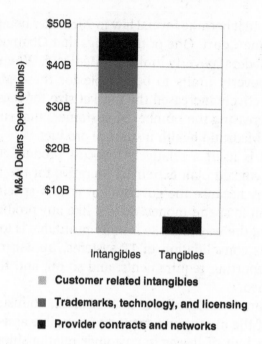

Figure 16.3

experienced revenue growth of approximately 700 percent within this given time period. All other Medicaid acquisitions totaled six, and there was a much lower amount of revenue growth.

To summarize: Intangible assets always matter the most (Figure 16.3). This is an example of a secondary transaction where the money was going to shareholders of the assets being purchased. It is an M&A example, and it is an example of how regulation impacted which intangible assets matter.

The Affordable Care Act was a massive reform in US healthcare, and the level of money spent was immense. The expansion of the eligibility based on expanding the percentage of the FPL available to new people that were now eligible for Medicaid opened up the market, and it gave companies the opportunity to grow through acquisition and particularly leverage cost synergies in that acquisition. It was because of the regulatory change that the customer relationship of a Medicaid consumer became more valuable.

Part II

Systems Engineering
Tools and Examples

Chapter 17
Systems Engineering Tools

Systems engineering work attracts graduates from a wide variety of degree programs. It is important to understand, though, that systems engineering tools are used differently from other professional tools gained by studying architecture, art, engineering, science, business, or mathematics. Systems engineering tools are fundamentally technical, and they are designed to take specific measurements through a myriad of processes and combine them into a single framework and language for mutual understanding across contributing teams. These tools will never be a particularly valuable crosswalk for connecting two non-technical, non-specific business units such as marketing and design.

The marketing and design departments can work together based on branding exercises. They can work together based on psychology. They can work together based on highly detailed customer analysis. But there are very few use cases where marketing and design can be brought together using the tools we have been discussing because systems engineering tools need highly specific inputs based on measurable data.

Since the dot-com boom and the rise of the Internet, the amount of digital activity that human beings encounter and the number of measurements and possible quantification of data that exists based on those encounters has increased exponentially.

The reason that systems engineering is now so valuable when it comes to intangible assets began its life in the dot-com boom in 1995 and the rise of the Internet. The amount of digital activity that human beings encountered and the appetite for measurement and quantification based on those technological encounters has increased exponentially. Prior to human beings interacting via a computer or any kind of digital interface, it would have been very difficult for a systems engineer to perform the kinds of analysis that they needed to fully utilize the functions of their tools. That has changed in the age of technology.

That is the fundamental element that you should understand about these tools and what their utility is to get the maximum value from this book. They need specific data, and they cannot function with vague inputs, qualitative statements, or highly strategic information that will not fit into a systems engineering framework. This is not to suggest that the highly strategic or the highly qualitative exercises of the work are not valuable; they are extremely valuable, and they have a very important role in the success of a business, but they do not apply to these people.

We discussed the series of tools that the systems engineer must be able to utilize in order to accomplish their job function effectively, and we are now going to walk through those tools within very specific use cases.

The use cases will be:

1. An advertising technology company, specifically as it relates to customer data and some kind of customer product that requires a login.
2. A logistics solution that is driven by technology and leverages supplier and buyer data as a platform – an example of this might be Uber.

3. A traditional manufacturing company with some kind of industrial product, such as metal coating or a physical product.
4. A service-based business.

The tools that we are going to explore within this section of our discussion on systems engineering include:

1. The customer affinity process.
2. The context diagram and context matrices.
3. Use cases and behavioral diagrams.
4. Originating requirements.
5. Decision matrices.
6. Goal question metric analysis.
7. Analytical hierarchy processes.
8. Precedent matrices.
9. Failure mode and effect analysis.
10. Interface tracking.

The reason I picked these examples is that I believe that they are the most effective at describing the broad utility that systems engineering has inside businesses all over the world.

There will be plenty of opportunities to explore these tools individually, so I will not go into that right now, but the important piece about understanding these tools at this moment is that each of them forces a user or a systems engineer who uses the tool to complete an exercise at a level of specificity that no other tool requires.

Tools inside academic curricula as well as in business, are generally much too broad or much too specific. For example, software code is a language that helps a computer function to accomplish some kind of input–output goal, and the framework for how software code works can be as vague as the rules of grammar or the rules of linguistics that may govern a specific kind of language. This means that the general dichotomy of too-broad and too-specific exists in all types of education. Therefore, systems engineering is an excellent bridge between this gap between specificity and generality.

The tools are relatively simple, and they can be used by anybody without a systems engineering education, and they are similar to operational or business tools that people who have studied business administration or finance will be familiar with, but what they force the user do is to think about systems and think about interactions between systems differently hence my reasoning for presenting them to you within this context.

I chose these use-case examples because I believe that they are very broad. So much so that many businesses will find themselves in one of these categories encompassing e-commerce/ad-tech, logistics, or manufacturing. Manufacturing captures all physical products that exist. I selected an industrial use case of the manufacturing industry because industrial products are generally business-to-business. If we were to explore something like direct-to-consumer, we would end up falling into the same analysis frameworks as logistics and advertising. And then advertising itself is what governs most of the Internet and consumer economy today.

The next step is going to be digging into these specific examples and how a customer affinity process or any of the tools in the toolbox can and will impact a business owner or an investor who wants to understand the specifics of these industry and business-use cases.

Chapter 18
Customer Affinity and Context Diagrams

F irst, I am going to present the customer-affinity process as it relates to systems engineering, followed by the nuances of how this process can change based on the three differ-ent industry examples that are driven by clients with whom I have worked. The three examples I will use will be the media, industri-als, and logistics verticals.

It is important to remember that the customer-affinity process is fundamentally an input-driven process. The process requires the systems engineer, in this case, the business investor or executive, to collect inputs that customers have given. This assumes that these inputs are available and that some level of research has been done that reaches a high level of quality, which indicates that the inputs coming from the customer-affinity process are accurate.

Step 1 with the customer affinity process is to take these inputs – different statements by the customer, generally collected through a survey which can be as few as five and as many as five million – and to group them into overarching benefits that the customer is seeking.

In a rudimentary example of a building, one would consider that the customer is looking for a safe building, a sturdy building, or a building that has certain square footage and amenities. If it is a residential building, they would want certain kinds of fixtures or architectural components that might make it feel more luxurious. All of these would be unique categories that customer requirements or statements would feed into. So, it is fundamentally a customer-driven process.

Now you need to consider how the customers change in our three examples of media, logistics, and manufacturing. In the example of media, we are looking through the lens of a media company that makes money through advertising. We work with media companies all the time that are in every kind of media space you can imagine; they want to offer what advertisers will pay for.

An example that many people will be very familiar with can be something like Facebook, and Facebook has been criticized in the media for this. The customers of Facebook are advertising brands. Brands that want to show a Nike ad or an ad for a trophy at the right place and the right time. The product, the material that Facebook is selling, is the human users on the platform. It is the data of the users. The individual using Facebook is the product. The value proposition that Facebook brings its advertisers is it can give the advertiser more benefits than competitors of Facebook. Advertisers want ads to be placed in a timely manner. They want ads to be placed in a relevant manner. They do not want an ad for something that could be considered controversial showing up next to an ad for their own product, and they want to be able to control the viewing structure of the ad to the customer. It all comes down to the ability to both serve the ad and to know factually that the customer data is correct.

So, the types of customer-affinity groupings that would come from the media company include accurate data, a user face to serve this data, real-time reporting, or reporting that can come back to the advertiser synch as did users click on it, did users not click on it? Juxtapose that now to the logistics example. We are talking about a company that gets a physical product from point A to point B. This physical product can be a human, or it can be a package. In those cases, the complexity of the customer requirements and customer affinities are actually very simple.

They want it to be faster, better, and cheaper than the groups that they are competing with or the alternatives that they have in the market. The customer-affinity process and logistics start to become highly specific based on the kinds of logistics business that you are talking about. Uber is a logistics company. The customer is ultimately the rider, the one whose credit card gets pinged when they order Uber to go across town and meet friends at a rock-climbing gym.

The other actor on this platform is the service provider which is the driver. The driver is the employee, the gig worker, or the contractor in this scenario. The driver has to be motivated to work on this platform so that the product can be available to customers. In the Uber example, customers want availability, convenience, cleanliness, and safety. But specifically, they want it within very specific territories. Very few people are looking for an Uber in farm country. The product – the driver, in this case – is not the customer. They would not be part of the customer-affinity process that one would conduct to group these statements that are collected for the output of the process.

The third example is the manufacturing of a physical product or a physical good for an industrial company. Think metal screws, electronics, and so on. These examples have business customers, and business customers have very narrow but deep requirements. Business customers need a very specific screw that is designed to very specific specifications, available in a very specific window of time that is updated based on any other change in requirements that might exist for the ultimate utility of the screw.

So advertising, logistics, and manufacturing will all have different kinds of customer-affinity groupings when you consider who the exact customers are. Customers may disagree with each other, but the customer affinity process can be a very useful tool for

early-stage companies that want to clearly define the problem that
they are ultimately solving and help customers articulate the benefits that they seek for that problem.

There are studies that have been conducted which indicate that
customers may say they want one thing but actually want something else. This is very common in taste testing for coffee, wine, and
alcoholic beverages. Customers claim that they want something
that tastes smooth and silky, but in fact, when they do a blind taste
test, they pick something that most people define as bitter or sour.
This is often done for the emotions that people associate with different words.

The customer-affinity process is something that has to be part
of either a new product development inside of an organization or a
new/startup company.

The customer-affinity process is simply about organizing customer statements that are impulsive or unstructured into at least
two levels of aggregation. And what I mean by two levels of aggregation is that if you have ten customer statements, then those ten
must be segregated into at least three different groups. On average,
you would have three raw customer statements in three different
groups. Those three different groups must be segregated into at
least two different groups (Figure 18.1).

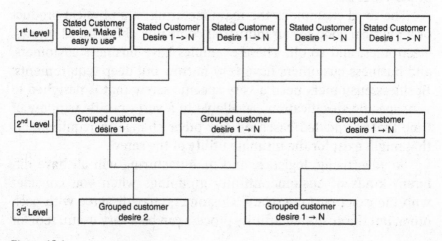

Figure 18.1

You start with a certain number of requirements. It would be too broad to group ten, twenty, thirty, or even five requirements or statements by customers into a single grouping. They will always be somewhat different. And it is the job of the systems engineer to pull out the specificity that exists within each requirement and then interpret that specificity into a subgrouping.

The exercise in the customer-affinity process is about getting between the first level – which I have indicated here on the chart – to the second level at a level of specificity that helps the engineer develop a perspective of the system that is more detailed than they started with. Beyond that, the only inputs to the system are raw customer inputs or data and then the engineer's own analytical thinking and logical mind. The customer-affinity process is not meant to be highly sophisticated. It is meant to force the engineer to think about what customers say they want and then articulate those wants, desires, and needs into functions and attributes that the system can have. For example, if you have a customer-affinity process for a laptop computer, and customers are talking about speed, load time, battery life, and turning it on and off quickly between classes, all of those may get grouped into a startup time set of requirements or customer desires, which would be the second level, and then the third level would be the BIOS (basic input/output system) of the hardware machine and the protocols that the firmware is running in order to expedite a startup time for a machine.

The second example that I want to talk about is the context diagram and the context matrix. The context diagram or context matrix is a much more useful tool for a mid-stage company. It shows how internal systems interact with external systems. So, a simple context diagram, such as the one displayed in Figure 18.2, is a flowchart of sorts where the center of the diagram is your system, the system that is being studied. All of the work, thought process, and analysis are driven to external parties. This is very important; the reason that I wanted to bring these tools together is that this is a very different mentality approach and psychology from the previous tool we talked about, the customer-affinity process.

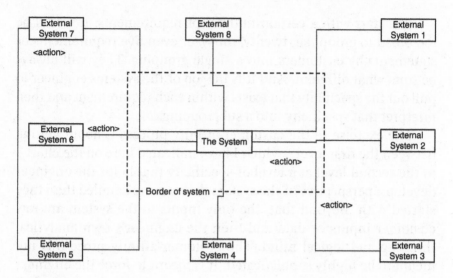

Figure 18.2

The center of the diagram will be your system, and the external system will be interacting with you. The kinds of external systems that interact with you and the ways that they interact with you are very specific based on your use case. Let us consider the use cases driven by the clients that I have told you about. In advertising, if we stick with our Facebook example, one of the external parties they have been having to deal with very aggressively since the Cambridge Analytica scandal is the government and its regulatory demands. Congress wants to ask questions to understand more about how Facebook uses its data, to whom it sells its data, and what Facebook knows about the people to whom it sells its data. Advertising is generally a low-regulated market, and ever since the Cambridge Analytica scandal, Congress in the United States has become more focused on the role that regulation should or should not perform inside the business model of advertising companies like Facebook. Regulation would not have been an external entity that we would have identified six years ago, but today, at the writing of this book in 2023, it certainly is, and it has been since the 2016 election.

The customers can be external, although the context diagram is not meant to just recreate some kind of broad analysis like Porter's

Five Forces (the number and power of a company's competitive rivals, potential new market entrants, suppliers, customers, and substitute products that influence a company's profitability).

A large external driver of advertising, particularly digital advertising, is the technology and the tech platforms that are available to collect this data from customers as well as serve this data back to advertisers in a compliant manner. The technology that targeted advertising existed before Facebook, but since people started to spend most of their time and their lives on social media, technology played an increasingly important role. Obviously, it is not Facebook's internal technology we are focused on in this example, but it is the external technology of different software platforms that may interact with Facebook's application processing interface (API) and Facebook login functionality.

Moving on to the logistics example, one could debate whether the drivers are internal or external for the purpose of this exercise, and for the purpose of a platform, I would say that the drivers on the Uber platform are external. They are not full-time employees of Uber for the most part, and that creates some very powerful dynamics between the drivers as well as Uber. They could be considered suppliers in the Uber example. Different impacts could be something as simple as the weather. If you are in the middle of a snowstorm or an area that suffers from inclement weather regularly, it is going to create a different utility or a different consideration for your system than if you are in beautiful southern California, where it is nice all year round.

The manufacturing business is significant, particularly in the case of the screw that we were talking about earlier or the widget. A manufacturing company has to have inputs and raw goods and materials, a different kind of supply. During the pandemic, when supply chains got constrained, the price of wax – which is generated from oil, from petrochemicals – went up, and candles became very expensive. Candles were things people generally bought impulsively inside a home goods store and used to create a more comfortable environment in their homes. When the price of wax went up for candle manufacturing, churches, many households, and individuals were suddenly ordering fewer candles as they were unable to afford them. This is just another instance of where the global supply chain and its ripple effects can impact something as simple and mundane as candle manufacturing, which prior to the shutdown, was a relatively simple activity.

So, the customer context diagram/matrix is an external view of how a business interacts with its other parties. The context that I coupled with the customer-affinity process provides two important and very different lenses of how to consider your system, both interacting with its own customers in the case of the affinity process and then interacting with different external stakeholders. It should be part of every business's roadmap, and it should be something that is kept in mind whether it is delivered as a formal deliverable or not. It should be in the minds of business owners, executives, and investors at all times.

Porter's Five Forces is a very common tool in MBA programs in business academia. It is the most similar tool to the context diagram/matrix, but if you are using an MBA lens, what I personally like about the context diagram and the context matrix within the perspective of systems engineering is that it is more than just the five forces. For each of the interactions between the external party and the central system, you have to identify the action that the external party is having on your internal system. In the example of the Congressional hearings and Facebook's data privacy policies, the professional hearings were increasing the costs for Facebook's privacy policies, and Facebook did wind up getting fined for that to the sum of billions of dollars.

Chapter 19

Context Diagrams and Context Matrices

The context diagram and context matrix are tools that can help us understand components that interact with a chosen system. For example, a context diagram might show the relationship between users, organizations, and systems that interact with a given business product. The context matrix takes those relationships and attempts to forecast their aggregated influence on output performance. What is important about using these two tools in unison is that they require a married level of specificity between the system one is analyzing and the subsystem. For example, a bad use of a context diagram and context matrix would be to have the internal system as an unmanned aircraft and then to have the external system as global aircraft regulations. "Global aircraft regulations" is too broad as compared to "an unmanned aircraft," which is the main system.

The core system and the external system are meant to identify what factors most significantly impact the use of a system. The context diagram in Figure 19.1 shows all the relationships that can influence the play experience of a toy called Little Chefs. The nature of the child playing, of other children in the room, the availability

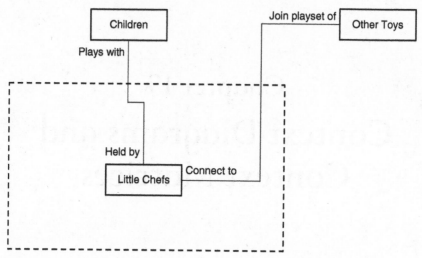

Note: simplified figures provided for learning purposes, final figures should contain more detail

Figure 19.1

of consistent, accessible electricity, the other toys available, the proximity of edible food or real life, serviceable cooking utensils, whether there is liquid nearby that could spill on the toy, right down to the material of the flooring under the toy – each of these has possible effects on one another or on the function of the product itself.

A context diagram can be a useful tool when dealing with corporate valuations. Imagine that a company wants to understand an intangible asset – a weather module that exists inside a consumer product application, like real-time weather radar connectivity for your fitness tracker. The subsystems and related systems that interact with that weather module, such as the hardware, the device, other software modules, the user, and social sharing features, are all tools that can help an executive understand better the kinds of value and potential strengths and weaknesses when representing the value of this module to investors or customers.

The context diagram is a tool specifically designed to show how external systems interact with your chosen system. It is usually represented by boxes with lines. The center of the context diagram includes your chosen system that has a broader outline with a dotted line to show us a level of focus on the system in subject. Your system, or the system being analyzed, should always be in the center of the context diagram. There are no limits or minimum requirements of external systems represented here by the box to the up, left, right, or in diagonal phases to represent. There is no limit. You can have as many of those as you want. It is best to have one line between the external system and the internal system because that will allow you to describe the interaction in the most simplistic way. This is meant to be a very macro or high-level view of how external systems impact your chosen system (Figure 19.2).

A completed diagram that might be applicable for an autonomous driving system is what I have chosen here in my context diagram example indicates that you have sunlight affecting the visibility of the pedestrians. You will notice here that the proximity of sunlight and the label "affects the visibility of" elements that are closely connected. The visibility is closer to sunlight, and that shows how sunlight is affecting the visibility of pedestrians. Then, you have additional systems such as planes, roads, trains, other drivers, cargo, pedestrians, or animals that might be on the road. You can grow this system into a larger system by showing how the sunlight affects the

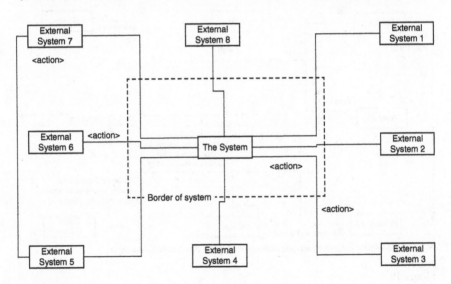

Figure 19.2

visibility of the pedestrians, which, in turn, can impact the system as well as how the other drivers can collide with, cut off, pass, or alert the autonomous driving vehicle with actions or intentions, and how that autonomous driving vehicle – within the expanded box – can provide some similar level of alert or interaction to other drivers.

The point of the chart in Figure 19.3 is to show macro events that the system is engaging in that are relevant to the receiver of the information. It is dynamic, meaning that the bordered outline box can change, but overall, it is meant to be very simplistic and can be extremely useful for software engineers, entrepreneurs, investors, and business owners to visualize a landscape of detailed actions, elements, vulnerabilities, or impediments to be contextually relevant to all parties.

The most important utility of the context diagram and what I want you to take away in this example is the level of specificity that exists between each of the different external systems with your own internal system. You will notice planes, animals, pedestrians, cargo, other drivers, and so on are all at a similar level of specificity. An incorrect level of specificity would be to – instead of talking about planes – add the

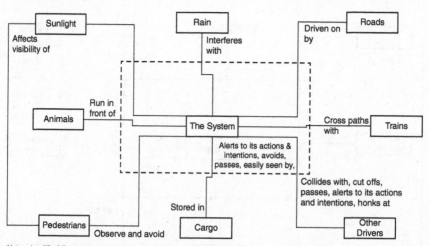

Note: simplified figures provided for learning purposes, final figures should contain more detail

Figure 19.3

landing gear of the plane as an element, and then, instead of considering other drivers talking about a fleet of other drivers. The landing gear of one plane is extremely specific, a fleet of drivers is extremely broad, and those two elements do not mix together, which is why it would be a poor use of the context diagram. The only instance where those two varying levels of specificity would be useful is if, for some reason, your system only requires landing gear and fleets of cars. Think about a military operation where the landing gear on a plane has some kind of defense capability that can neutralize a threat that is coming from a school of cars. In that case, the example would be appropriate. In this case, since we are talking about an autonomous driving system, it is inappropriate because the example is too extreme.

Chapter 20

Use Cases and Behavioral Diagrams

A nother useful tool in systems engineering that executives can use to understand which intangible assets matter the most is called a use case and behavioral diagram. Use cases refer to the basic functions that a customer or stakeholder expects a system to perform. You might purchase a toy expecting it to provide some level of constructive playtime, and maybe there is a light feature, vocal, or musical component. The value of the feature is in how the user experiences it. We call these basic expectations the use cases. Unsurprisingly, the use cases that matter the most are generally the ones that matter the most to the user.

A behavioral diagram (see Figures 20.1 and 20.2) helps us understand how a use case, an initial condition, interacts with an operator's decision, resulting in a new exit condition. So, for example, if a child is able to change the outfits of the Little Chef, and several children play within this same space, and each child changes the state of the toy, the toy will find its way to a new exit condition.

Use Cases
Little Chefs can be handled by a 4-year-old child
Little Chef uniforms are machine washable

Figure 20.1

Behavioral Diagram – Illustrative		
Initial condition	Little Chef may change uniforms	
	Little Chef newly purchased	
	Operator: Child	Little Chef
	Child selects uniform for Little Chef	
		Little Chef affixes to uniform
Exit condition	Child places uniform onto Little Chef	

Figure 20.2

Diving deeper, system use cases essentially describe how a system user, actor, or stakeholder uses the system.

Use cases are not written in a specific way like requirements. Use cases can be as simple as "child opens door." The four use cases I have selected for this example pertain to the user logging into a system. Use cases are meant to be written in a way that communicates the informality of how users and stakeholders interact with the system.

A behavioral diagram provided for an illustrative login system can be as detailed or as broad as you want it to be. I have selected an initial condition where a user logs in with a unique combination of a password and username and an exit condition where a user has selected this unique password name that is saved in the system. All I have identified in this behavioral diagram is the user creating their own password and username combinations. You could have more rows than this. It would depend on the scope of the systems engineer in the system (Figures 20.3 and 20.4).

You could very easily go down to the detail of the SQL database saving this information, querying it, and so on. You could even go as detailed as determining the number of character strings and requirements that a password may have.

This example is provided illustratively to make sure that you, the reader, understand how this basic string of text can work in a behavioral diagram. The behavioral diagram is intentionally meant to replicate the visualization of loops of code. Code programmed in modern languages often looks similar to this with the same

Use Cases
The system opens at 9am NYC time Monday through Friday
Users login with a unique combination of password and usernames
Users save contact information to their online profile
The system may change its outer appearance

Figure 20.3

Behavioral Diagram – Illustrative			
Initial condition	Users login with a unique combination of password and usernames		
	User creates new account		
	Operator: System	Password Database	User
	Users type in unique password and username combination		
		System performs query to determine if combination already exists	
			User receives messaging confirming combination is accepted and successful
		System saves information	
Exit condition	User has selected unique password and username combination that is saved in system		

Figure 20.4

indentation structure, as you have loops and sub-loops that perform different functionalities within the different hierarchies.

Use cases and behavioral diagrams force a business owner to consider the way in which a system can be used and can include something as simple as "the user logs into the system." Or "The user logs out of the system." Use cases are very popular for software development.

An important dynamic of use case drafting that is different from requirements drafting is that for requirements drafting, the verbs used are definitive. "This system shall provide a login interface. This system will log the user out when they click the logout button." They need to be definitive to do their work.

In the examples of use cases, the phrases are structured to be more descriptive and less absolute to walk people through how a system can be used. Use cases are deeper dives into some of the verb relationships we were talking about within the context diagram earlier. I mentioned the updated reform policies which increased the costs or resulted in fines for Facebook's compliance department. It could be something as simple as "Congress reviews policies of Facebook." That is a simple but important example of a use case. That would not necessarily be a use case of the system; that would be a use case of the context diagram, which includes these external players, but it serves the point of understanding how simple the phrase can be.

The purpose of the use case is to break down in simple sentences what a business or system does. In technology, and particularly for smaller companies, I often hear an excessive number of buzzwords where entrepreneurs are trying to capture recent headlines that they read in TechCrunch, *The Wall Street Journal*, *The New York Times*, and so on. That obsession with buzzwords – which is driven by entrepreneurs as well as investors – often dilutes the basic functions and simplicity of describing what a system does, which ultimately makes it more difficult to understand.

The most important takeaway for this use case for the business owner, investor, or executive, is that the use cases should be drafted internally within the mind's eye of these three stakeholders to very simply put forth to any interested party, what a system does.

Use cases for media companies can include something as simple as "The system provides real-time ad displays. The advertiser uploads desired advertisements. The advertiser receives reports based on the effectiveness of the campaign." It makes it clear to understand who is doing what in the system and the value that is being received by different system stakeholders.

Use cases in the example of a logistics company are "User books a ride. The user enters the vehicle, the user uploads their credit card." Something that is more enterprise-driven would be a "The business downloads a shipping label. The business sends packages to customers." That simplicity is required for all stakeholders to clearly understand the process. On the manufacturing side, the example is "Company produces screws. Screws measure six centimeters long." It is that simple.

The number of use cases that can exist within these systems, any minute subsystems or derivatives of the systems, can be extremely large, but the simplicity is maintained.

The use case utility is very different from the tools that we have reviewed up until now, which are the customer affinity process and the context diagrams, because the use cases are highly specific and pedantic. They force the stakeholder that is utilizing the tool to make it simple. If use cases are not created with simplicity at their core, it will be to the detriment of all parties.

Chapter 21
Originating Requirements

Originating requirements are an evolution of use cases. They describe what a system should do on a factual and definitive basis, such that it is a requirement that the system performed the desired function. Otherwise, the system would be determined in a failure state. The most important technical component of requirements is that they use specific verbs, and those verbs most commonly are will or shall (Figure 21.1). A computer shall weigh ten pounds. An operator will log in at 8 am New York City time. These are requirements. They must happen in order for the system to function. This is different from use cases where – juxtaposed to a use case – the originating requirement is more definitive, and the use case is more fluid. The comparative use case in the two examples of requirements I just gave you would be considered something like "the operator will arrive at work in the morning."

	Originating Requirements	Abstract Function Name
A1	X Shall/Will Y	Startup

Figure 21.1

Originating requirements describe a system and how it functions. They are the backbone of many systems engineering processes, and they identify how the product is expected to work down to a very specific level.

Originating requirements are not a marketing or sales statement. The best way to describe their purpose is to consider how they would be presented to a project manager. Imagine if, at any point, a project was handed off to a contractor or a vendor who was commissioned to complete the project; the contractor would need to be bound by these requirements lists.

The requirements become limits of the system, and there is a defining line between interpretation and what the designer intended. Requirements need to be specific, but not so specific that it gets down to a level of "Press A on the keyboard, press P on the keyboard, spell the word apple on the keyboard." That would be far too pedantic. In their entirety, originating requirements represent how a system functions as a whole while defining its unique characteristics. Originating requirements are very popular in software engineering because software code is definitive and forces inputs and outputs to come together into an amalgamated algorithm that accomplishes a function.

They are also very common in technology and are usually hotly debated among engineers. Businesspeople do not care about originating requirements. The value of an understanding of originating requirements for an investor, an executive, or a business owner is to consider how some originating requirements would be applicable based on a given example.

There can be functions assigned to originating requirements. You can have functions like accessories, interfaces, collaboration and design, and changeability, and all of those would help you to club or group originating requirements so that they are easier to collate. These categorizations are meant to communicate the general function that the requirements affect, including material function.

Let us consider now the three examples of originating requirements. This example of using the originating requirements with advertising technology, logistics, or manufacturing has to be more specific because it would be difficult to assign an originating requirement to the business level. Originating requirements must be extrapolated to the system or subsystem level, as is very often the case in software code. For example, consider advertising technology and the ability of an advertising platform to serve an ad when a human is in a specific location. This is done very commonly in different digital out-of-home businesses that utilize geofencing or geolocation technology. For example, the screen shall display an advertisement for Adidas when more than 30 percent of the humans are determined to be potential Adidas buyers within a ten feet radius. This is a good example of an originating requirement.

It is important to understand that originating requirements are the highest level of requirements in themselves as a hierarchy. Below originating requirements, you can have derived requirements. In the instance of a derived requirement for the geolocation example I just gave you, you would need to consider perhaps that there is a camera. "There shall be a camera on the digital billboard to determine the total number of possible viewers within any ten-second period." The derived requirement would inherently be defining the system's need to determine when more than 30 percent of the proximate viewers are our potential Adidas customers.

Considering logistics like Uber, FedEx, or DHL, the originating requirements will be numerous. They could be as simple as "a user logs on to the Uber platform," or it could be as complex as "platform pricing engine shall display surge pricing when there are more than ten people looking for a pickup and drop off within one mile of each other" – or "pricing engine shall display surge pricing when it is raining or when precipitation reaches a prescribed level."

Lastly, for the manufacturing of a physical product or industrial good, these requirements can often be physical material or hardware driven, "the screws shall withstand total pressure of over 10,000 pounds." A screw that withstands 10,000 pounds is different from a screw that withstands 1,000 pounds, and it is different from a screw that withstands a 100,000 pounds, and that is the value of the originating requirement as it takes generally vague use cases and generally qualitative statements, no matter how much the systems engineering tools have been employed, and it makes them extremely specific. "This screw will withstand 10,000 pounds therefore, it must be made from a specific material, tempered by a specific process, and so on." All this information is relevant to me as a systems engineer.

Originating requirements help chisel away at the raw marble to create a beautiful sculpture that ultimately becomes the system. You can have a screw that withstands a pressure of 10,000 pounds and simultaneously is one inch long and a quarter of an inch wide and simultaneously weighs less than a certain number of ounces or grams. All of those originating requirements could come together to describe a very specific screw. There is no other highly effective tool in systems engineering or in most pedagogies that allow you to describe such a specific solution as the required screw's characteristics.

Originating requirements should be utilized by managers, business owners, and decision-makers because they allow owners and executives to be highly specific.

The role of drafting, integrating, and monitoring the originating requirements and derived requirements is not up to the decision-maker. It should not be the role of the business owner, investor, or executive. That should be delegated to the software engineer. But understanding and utilizing this level of specificity, even in boardroom conversations, will help business owners, executives, and investors have a more substantive conversation.

Delving a little deeper into the functionality of and interactivity between the originating requirements, activity table (Figure 21.2), and precedence matrix (Figure 21.3) tools let us look at some real-world examples. As is the case with most of the tools in the systems engineering suite, originating requirements are reasonably simple. They are numbered with a letter and a number, so A1, A2, A3, B1, B2, B3, and so on, and the letters identify what category the requirement belongs to. Requirements as a tool are very simple, but their utility and their function are very high.

	Originating Requirements
1	The system shall launch a projectile
1.1	The projectile shall weigh less than 5 kgs
1.2	The projectile shall be launched between 3 and 6 meters
1.3	The projectile shall be made of rubber
2	The system shall store energy

Figure 21.2

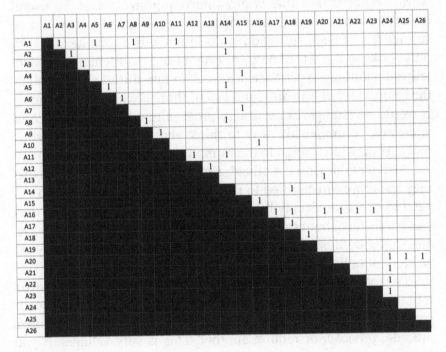

Figure 21.3

The purpose of a requirement is to state an action, attribute, or characteristic that a system must have. It is not optional for a system to have a requirement. A system must meet requirements in order to achieve its objective. To juxtapose two examples – activities/use cases versus requirements: in a use case example, we talk about "user logs on to system and reviews reports," versus a requirement which might be "user shall have single log on credentials, user shall be able to view reports after logging in," or, "reports shall show purchase history over the last twelve months." The reason requirements are more specific and limiting than activities and use cases is that requirements always use the verbs shall or will. Every single requirement must have the verb shall or will. So, X shall Y. W shall V. Z shall D, as in Figure 21.1.

The purpose of using the terminology of shall or will is to make it definitive, not "the user logs in and dances around different functions and then logs out." "The user will have a login of ID and password." What makes this originating requirement exercise useful is that human beings do not speak in requirements. It would be strange if you were sitting down to lunch with someone and they only spoke in the terminology of originating requirements. "I shall tell you about my experience last week in jury duty. I shall recollect the funny story of jury duty, and I shall share it with you right now." It would be strange. So, it is very programmatic. It is a very zeros and ones, computer-like way to think about a system interacting that is somewhat foreign to the human cognitive functions.

Activity tables, on the other hand, are like use cases, except that they are more directly tied to requirements (Figure 21.2). I do not need to delve too deeply into activity tables other than to say they essentially talk about the major activities that a system should have, and that should be closely related to the requirements, but they are written more like use cases.

The reason activity tables are relevant for the precedence matrix is that in the precedence matrix, you tabulate all of your use cases or your activities, and then you number them and identify which ones precede, proceed, or require another. This is very valuable if you come up with a thousand originating requirements, and then on top of that, you have 10,000 use cases or 10,000 activities. Each of those activities is either in parallel to, before, or after one another. That has to be diagrammed, and that is the purpose of the precedence matrix. To look at the set of activities in use cases within a system, not at a coding level, but at a functional or almost an operational level, and then identify which activities proceed which.

Figure 21.3 shows what a precedence matrix looks like when it is completely developed – more on this in the next chapter.

Chapter 22

Decision Matrix

The decision matrix follows the goal question metric (GQM), and it is used to rank the opinions that are developed about the performance or the relative value of a solution or a goal. Take, for example, the toy. If we are dealing with a toy that has various attachable accessories, we would need to evaluate the method of attaching those accessories. We are going to need to choose between, for example, clips, Velcro, bendable hands, sticky surfaces, magnets, and the like. An objective scale would be used to analyze each option across the most important dimensions like reliability, creativity, cost, diversity of use, and so on. The mathematics of this can be seen in Figures 22.1 and 22.2.

FINAL SCORE				
Clip	Velcro	Bendable Hands	Sticky Surfaces	Magnets
3.091	2.750	1.274	3.25	3.429

Figure 22.1

Score						
	Weight	Clips	Velcro	Bendable Hands	Sticky Surfaces	Magnets
Reliability	1	0.727	0.500	0.364	0.667	1.143
Creativity	1	0.091	0.250	0.273	0.333	0.214
Cost	5	1.364	1.250	0.273	2.083	1.786
Diversity of Use	2	0.909	0.750	0.364	0.167	0.286

Figure 22.2

You can look at the relative weight of the four dimensions – in this case, reliability, creativity, cost, and diversity of use – based on the mathematical determination of how each of them was scored within a ratio of one to hundred (Figure 22.2). It is very simple to perform this analysis because it shows how the interrelation of the different variables can be used to rank them against each other and ultimately create a score that adds up to 1 percent or 100 percent.

The method used for this decision matrix involves objectively determining how the variables selected match up against criteria, then taking the weighted average of each option's score and the total score for that option category. After this, you can multiply the standardized weight of each of the ranking values to determine the score for that option. Finally, you sum the scores for each category of each option to develop the final scores given in Figure 22.2.

The decision matrix is very often used in human capital to rank options of a developed concept based on an engineer's or a user's own perception of the value of those options. For example, you can take several different components of a physical product – perhaps the durability of a physical product, the number of buttons it has, the amount of material that a physical product has – and then you can weight them in regard to how they tie into the goals that you created, perhaps along the categories of the reliable, creative, cost-effective, and diverse. Or SMART goals which are Specific, Measurable, Achievable, Relevant, and Time-Bound.

It is important to remember that the decision matrix has to tie into the goals that you have created. We went through a series of goals in our previous exercise that focused on those three different industrial use cases.

It becomes very difficult to separate the different methods that may be specific to a subsystem as they relate to goals. Particularly in the absence of a specific system that you are considering. So rather than including this graph in the appendix or in the document itself, I will allow the user to leverage the exercise and ultimately understand how these decision matrices can be influenced in the GQM analysis.

We are now going to review the figures for the GQM decision matrix and analytical hierarchy, all clubbed together. GQM is very simple. You outline the goals, the questions that you need to achieve the goal, the ideal metric, the approximate metric, and the data-collection method. The philosophy behind asking yourself these questions for how the goal will be achieved or how it will be measured ties back to a manufacturing and engineering principle of the "five whys." You must ask yourself why five times and have meaningfully different answers in order to really get to the root cause of what is happening in the system, or somebody can argue non-engineering scenarios.

The purpose is to elicit goals that are separate and that can be measured so that there is an idea of what the perfect metric is. It is very rare that engineers will have the opportunity to select the perfect or ideal metric for measuring the accomplishment of their goals, but they should always have an understanding of what that ideal metric is because then it allows them to select a more effective or appropriate approximated metric.

The decision matrix and analytical hierarchy is simply a method to rank the goals that you have developed based on a relatively objective weighing scale by creating subjective weights. First, you define the number of goals that you want to weigh against each other. You must always have a plural number of goals and subgoals.

Therefore, you cannot weigh one goal against itself. You cannot weigh one subgoal against one parent goal. You must have at least two goals, and each of those two goals must have at least two subgoals. In this example, I have chosen two goals with three subgoals each for a total of eight measurable and weighable goals.

You simply define your goal, then you define your subgoals, and you apply a percentage weight of importance. It is important that the goals are defined within the same scale, meaning you do not want to have one goal saying, "be happy" and then another goal saying, "not be sad," because a high percentage of not being sad is the same thing as a high percentage of being happy. Those two goals will end up conflicting with each other when you try to run out the linear program. In step 1, you define your variables (Figure 22.3).

In this example, I have just selected N and M. Goal 1 is N, and goal 2 is M. Subgoal 1 of M is M_1, M_2, M_3, and so on. You then assign weights. It is very important that these weights add up to 100 percent for each level in the hierarchy. Then, you simply multiply each

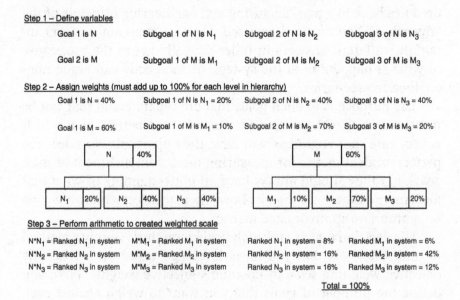

Figure 22.3

subgoal by its period goal to come up with the weighted scale of this goal inside the system that you have developed. Using basic multiplication, we were able to develop a consolidated ranking for each of the N_1, N_2, and N_3, and M_1, M_2, and M_3 in relation to each other, and you will notice that in all cases, the totals of the weights add up to 100 percent. This then allows you to say mathematically and definitively that M_2, which is ranked at 42 percent, as you can see from the previous arithmetic we did, M_2 is 70 percent. Its parent goal of M was rated at 60 percent. Seventy percent of 60 percent is 42 percent. M_2 became ranked the highest in this selection of six subgoals that we are measuring. We have also mathematically determined that N_2 is twice as important as N_1 based on the relative rankings of our system.

The reason this tool is fascinating and so useful is that there is truly an infinite number of goals and subgoals that you can develop (Figure 22.4). You can develop a decision tree in a matrix that rolls all the way up to a top-level goal, which then ultimately allows you

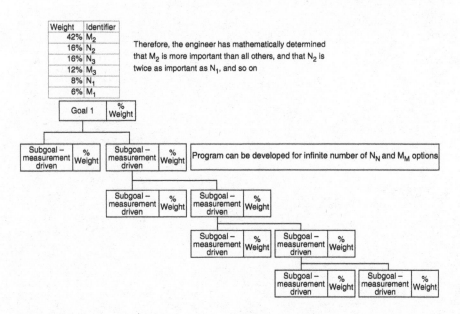

Figure 22.4

to assign the total weight of the goal that is at the parent level. So, for instance, in the case of N, the relative ranking of N_1, N_2, and N_3 is 8, 16, and 16. So that means that the aggregate weight of N, the parent is 8 plus 16 plus 16, which is 40 percent and has been confirmed with the mathematics that we exercised.

Chapter 23

GQM – Goal Question Metric Analysis

The goal question metric (GQM) approach is a systems engineering tool that is used to create and rank goals, questions, and metrics based on a standardized methodology that is relative to the task at hand. GQM analysis is used in software development, but it is also useful for hiring personnel. This explains why it is popular in the human capital industry when trying to create a set of criteria that will help weigh the strengths and weaknesses of candidates. I will repeat the usage of some graphics here from our discussion on decision matrices.

Much of the work done during GQM becomes a springboard for decision-making around the costly development of different solutions that might rank highly on a GQM scheme. The GQM process is simple.

1. Start by listing the goals of your organization or your project.
2. Develop questions that serve as an assessment of whether you are attaining those goals.
3. Decide the metrics that will help you answer these questions.
4. Figure out how you are going to consistently collect the data you need to feed into your metrics.

Goals	Questions	Ideal Metric	Approximate Metric	Data Collection Method
Goal of system or subsystem	How will the goal be achieved? How will the goal be measured? 5 Whys	Assuming infinite resources to test	What is readily available to the engineer	Surveys, reports, devices, etc.

Figure 23.1

Goal	Questions	Metric	Data Collection Method
Maximize entertainment	How long are the toy timers?	Seconds	Measure
	How many accessories does the toy come with?	Number of Accessories	Count
	How long does the child play with the toy?	Hour of playtime	Observe

Figure 23.2

Ideally, the data collection method is something that you will be able to test later on should you choose to employ these tools (Figure 23.1). The matrix in Figure 23.2 shows an example of the GQM approach, with the overarching goal of developing an entertaining child's toy.

What are the goals of your current project? What kind of questions might you develop that would allow you to measure this goal? For each question, what is the most useful unit of measure, and how will you collect that measurement? Are you measuring the weight of a precious metal extracted from a mine or one used in a product, for example?

You might collect data in grams of gold per ton of ore mined, while the ultimate performance value of the mine may be communicated by the total cost per unit extracted. A little time spent walking through this simple process will help you understand and communicate clearly with your team precisely what your goals are, what

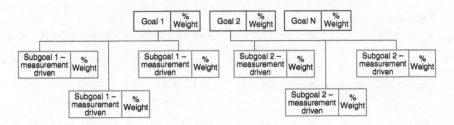

Figure 23.3

defines success, and the most appropriate questions to help measure and communicate your progress along the way (Figure 23.3).

There are a lot of components of GQM analysis that allow managers and investors to break down intangible assets. Remember, an intangible asset is fundamentally something that cannot be felt or touched, but at some point, it must succumb to measurable criteria. Select, for example, a business owner that makes luxury watches with the explicit goal of developing the intangible asset of "recognition" among the most premium luxury brands. What are the questions that would allow this business owner to analyze this goal? How about the amount of money customers are willing to pay for his products? What types of people will wear them? Is there a way to measure the overall measurable quality of the watch? Is the watch cosmetically flawless? Can we internally grade the precision, fit, and finish? What about the number of components that make up the movement? How long is the timepiece expected to last? What is the aggregate value of the precious metals and stones utilized in the design (Figure 23.4)?

All of these questions point to the ultimate goal of recognition as a luxury brand. With these questions clarified, the measurements become self-evident:

- Volume or weight of precious metals.
- The average life cycle.
- The average duration of time before the watch is effectively retired or transferred to a new owner.
- Income or net worth of the typical owner.

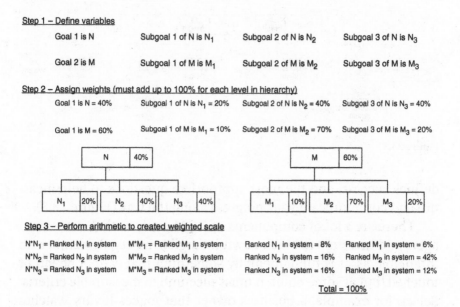

Figure 23.4

- The number of known celebrities who have purchased (modified by a factor of the average number of seats they sell at the box office or how many streams, downloads, or subscribers on YouTube or another platform).

Each of these metrics standards contributes in a unique way to address the question of whether or not this watch fits alongside other established luxury brands (Figure 23.5).

So, when a business owner says they want a premium or a luxury brand, what do they mean? Well, they want to sell the product for as much as possible. They will need celebrities of a certain status to wear it. They will have to meet or exceed the market's expectation of the incorporation of various precious stones and metals in the design. The business owner may want the watch to be priced at a minimum of ten thousand dollars, maybe even twenty, fifty, a hundred, two hundred, or five hundred thousand. But that is not what they say. That goal is not sufficiently specific to help plan and plot the progress of the brand. The overall overly simplistic goal of selling a watch for an exorbitant amount of money has to be broken

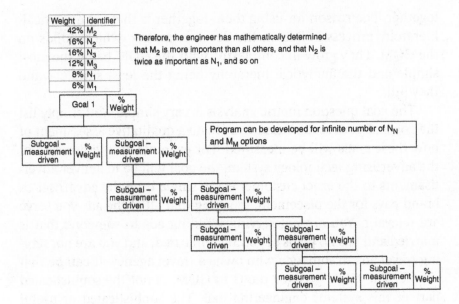

Figure 23.5

down into measurable, specific, attainable goals that yield simple analysis.

I will say it again, systems engineering provides an extremely powerful set of techniques and principles that business owners can use to understand intangible assets. The fact that GQM analysis has been popularized by systems engineering simply proves that the tool has high levels of utility inside complex engineering scenarios that would be otherwise difficult to define. But GQM analysis is also something you probably do without realizing it whenever you undertake a simple brainstorming exercise. Execute it with a little discipline and with consistent metrics over time, and you have built a strong case to establish the value of the underlying intangible asset.

There are three tools that are best used together. They are the goal question metric, a decision matrix, and then the analytical-hierarchy process. They are presented as three separate tools for you so that you can diversify the utility that the tools provide, but as I move through these three different examples, I will use them together. The reason for using them together is that the analytical-hierarchy process builds on the decision matrix, which builds on the GQM. They grow in complexity, with the GQM being the most simple and the analytical hierarchy being the least simple – and they link.

The goal question metric analysis is very simple. You simply list the goals of your system, and you create a qualitative assessment of how these goals will be measured. For example, in the use case of the advertising technology system, the goal will be to deliver advertisements to the exact customer every single time the advertiser or brand pays for the placement, meaning that all of the ads you serve are relevant. You are not serving wedding ads to someone that is married and has no plans of getting remarried, and you are not serving travel ads to someone who owns a travel agency. It can be that simple. The GQM goal, or the G in GQM, is not the sophisticated part of this systems engineering tool. The sophisticated or useful part comes in the Q and the M pieces, which is the question and the metric. This is very important. This subtle psychology about how goals are measured is something that is missing entirely from the curriculum in all management pedagogy that I have interacted with.

You have goals, for example, to increase profitability within a division or to grow more customers within a specific market. The way in which you measure those goals or those subgoals is never as specific as it should be. The metric has to be extremely specific and can only be as specific as the goal. So stepping away from our three use cases for a moment and using an anecdotal example, if I say that my goal is to be happier, which is a very vague and broad goal, then the metric could be as simple as I say I am happy. It is a self-reported metric. On the other hand, if I say my goal is to increase serotonin production by 15 percent between the hours of 6 a.m. and 8 a.m., five days a week, suddenly, the ideal metric becomes much more specific, and it becomes more specific because my goal is more specific. Imagine a business owner or executive saying, "I want to increase profits." The metric for that is obvious, but the goal is too broad. The goal has to be more specific. In the same anecdotal framework, the executive can say, "Between February and March of 2024, I want to increase the collection of profit for every sock sold inside the state of New Jersey that is direct-to-consumer." Suddenly it is a very different metric, and the goal is more specific but relatively the same.

It is important that business owners, executives, and investors do not deal in the world of abstract jargon, or what many people refer to as BS. Investors, business owners, and executives must hold themselves accountable to be specific because that is when true progress can be made. This is at the core of every single systems engineering tool that we talk about in this book.

Back to the GQM analysis. For advertisers, the goal has to be specific to the function of the business or the technology. In the case of serving an ad at exactly the right time and exactly the right place, the metric then becomes some level of measurable feedback loop to determine that this ad placement has achieved its goal. The last piece of the GQM analysis is the question. To force yourself to be specific, If an ad is going to be placed at the perfect time in the perfect place, the question cannot simply be the inverse or contrapositive of the goal; the question cannot be, "Was the ad placed at the perfect time in the perfect place?"

The question has to get to something deeper that the goal is trying to measure. There can be multiple questions for one goal. For example, if the goal is to serve the perfect talent, at the perfect time, in the perfect place, then the question would become how many ads were served within this time? How many ads were incorrect? How many ads were verified to be correct? How many ads were correct within a certain category? If we stick with the wedding example, how many ads were served to people that actually needed wedding accessories or wedding utility? It is that specificity that allows you to begin to measure truly what is going on inside your system.

With regard to the use case of logistics, goals are often about delivering things on time. If the person is ordering a ride through an Uber ride share program or if a person is shipping an e-commerce product from Shenzhen, China, to San Francisco, the question becomes related to shipping the product on time. But then the question becomes how you define "on time?" The question becomes, which endpoints are you talking about, do you mean time in terms of business days or calendar days? These kinds of questions help lead to the ideal metric, which is about the number of days from the order being placed to the order's arrival. An ideal metric could be the number of days gone since the customer received communication about an order, and so on.

The last example of the industrial product for the manufacturing business on a B2B scale, a common goal can be something like "deliver a physical product that meets certain specifications such as durability or just make a durable industrial product." Then, the obvious question becomes how you measure durability in terms of tensile, compressive, shear, torsional, or yield strength of the unit, or the materials that the unit is made with, and so on.

That is GQM analysis. It forces the business owner, investor, or executive to avoid the jargon and broad speak, ensuring they get real with what they actually want to change, as is very common with systems engineers and the way systems engineers work and think.

As a reminder, GQM analysis is to be used in partnership with the decision matrix and the analytical hierarchy. The analytical hierarchy is very mathematical, which we will get into in a moment, and it should be used as such.

Chapter 24
Analytical Hierarchy

T he analytical hierarchy process is a way to assign subjective weights to goals that have been identified in the GQM. Weights are divided up based on priority and calculated so that a weighted average can sum to one. We are then able to determine each goal's relative priority to other goals being considered. Figure 24.1 shows a chart of analytical hierarchy driven by the GQM analysis.

	Make the toy fun			Make the toy simple					Make the toy safe	Make the toy durable
1	50%			30%					10%	10%
2	Make the toy entertaining	Make the toy friendly looking	Make the toy playable for a long time	Make the toy easy to carry	Make the toy with few Components	Make the toy perform simple movements	Make the toy easy to assemble	Make the toy interactable with unrelated toys	Make the toy safe	Make the toy durable
3	40%	40%	20%	20%	20%	30%	10%	20%	100%	100%
4	20%	20%	10%	6%	6%	9%	3%	6%	10%	10%

Figure 24.1

You can see, in this case, we are using the toy example again. Across the top, you can see our parent goals from the initial GQM, listed in order of priority: making the toy fun, simple, safe, and durable. The total weights of those parent priorities add up to one. Written vertically, we have decided on some sub-priorities that move us toward those goals, giving each its own weight, with the sub-priorities in each parent category also adding up to a total value of one. Then, across the bottom, we simply multiplied the parent priority weight times the sub-priority weight. These values also add up to one.

Now we have an objective measure of how making the toy entertaining – 20 percent of the total priority for this toy – ranks against making it easy to carry (6 percent of the total priority). Making the toy look friendly is much more important than making it easy to assemble. This is an excellent tool for removing as much subjectivity and bias as possible when rank-ordering priorities. You now have a clear rank order of the relative priority of each of your product design goals.

The analytical-hierarchy process delivers subjective weights to the goals that have been defined inside the GQM. Weights are then divided up and prioritized so that they can be summed to one.

This is a weighted average exercise and is based on relative assignments that the user, systems engineer, business owner, executive, or investor assigns to each of the specific goals as they go along.

The reason that the subjective nature of this can work for the purpose of systems engineering is that the general biases of the engineer can be taken as fact or valuable. Meaning that if an engineer prefers quality over durability, then the user can assume that is the right order of preference. This can be done because the engineer is ultimately the owner and the developer of the system. There is some trust in the creator with these tools. These tools are not meant to be argumentative, litigious, or combative. They are meant to elicit a deeper level of thinking within the user or the creator.

Chapter 25
Precedence Matrix

The precedence matrix is driven by the originating requirements and derived requirements collections. It utilizes a crosswalk between these requirements and how they integrate with each other, particularly which subsystems they impact.

The precedence matrix shows how an originating requirement is related to other dependencies. Earlier, we were talking about how a screw must weigh a certain number of grams or ounces. This could be derived from the manufacturing process and the way that the screw was hardened and molded during that manufacturing process.

Dependencies are very important for originating requirements. You cannot have a skyscraper that is a hundred stories tall if you do not have a foundation that has a certain amount of mass or a certain amount of stability.

The words used in determining or articulating these dependencies are terms such as precedes, requires, or triggers. A skyscraper that is to be one hundred stories tall "requires" a foundation that is at least two stories deep. Words are associated with these requirements so that they can be accurate when describing these integrations.

	Controller Power	Cannon Power	Cannon Weight	Cannon Housing	Cannon Attachment	Cannon Belt
Controller Power						
Cannon Power						
Cannon Weight				Precedes		
Cannon Housing					Precedes	
Cannon Attachment						Requires
Cannon Belt						
Controller Weight						

Figure 25.1

The goal of the precedence matrix is to have an upper triangular matrix where words are placed that indicate how different systems interact with each other. In this upper matrix, it could create a loop communicating satisfied requirements that would be applicable based on a given integration between two systems. An example of this is displayed in Figure 25.1.

The diagram demonstrates how in this use case – which is a precedence matrix for a toy – the controller power, which is a subsystem, cannot interact with the controller power itself because that would be ridiculous. But the controller power can interact with the cannon power, the cannon weights, the cannon housing, the cannon catchment, or the cannon belts.

Let us now consider how a precedence matrix would interrelate subsystems. A precedence matrix is a systems engineering tool that brings together the context diagram, which was reviewed in Chapter 19, along with the originating requirements, which we reviewed in Chapter 21.

The context diagram, as I demonstrated in Chapter 19, describes how advertising technology requires many different systems, subsystems, and external systems.

Data interfaces, which we will explore a little later in Chapter 30, become a very important part of being able to serve an advertiser the right point of view of customer data that the advertiser needs to make good decisions.

Considering Facebook as an advertising company, an overall system could be a user's profile, and a subsystem could be the database that includes the profile edits that they have made over the one, five, ten, or fifteen-year history that the user has used the social media profile.

The precedence matrix in this example would show that the profile updates over the last decade is dependent on or becomes triggered by the overall system, which includes the user's general profile. Similarly, in logistics firms like Uber, FedEx, or DHL, subsystems can include routes, vendor systems interaction, rider interaction with the navigation subsystem, and so on, and all of those subsystems would have to be harmonized together to be consistent.

Then, in the case of the manufacturing of a physical product or good, the subsystems are more clearly shown in the product development life cycle, which includes items specific to the product, such as power source, colors, materials, durability, and other kinds of physical characteristics.

Chapter 26

Failure Mode and Effect Analysis

L et us consider the failure modes and effect analysis for all three of our use cases on the industrial side. The failure mode and effect analysis is used to evaluate risk factors that are associated with a system. It is all about failure. The FMEA diagram is all about identifying ways in which a system can fail or will fail based on a set of dependencies. It gives the systems engineer a chance to look at the inverse of the problem that it is trying to solve, which is usually along the happy path or the operational path of a system. What happens when the system does not work? How can you determine the ways in which it might fail? That is what the FMEA focuses on.

Consider the FMEA for a technology company: this can be as simple as a platform losing power. Obviously, all digital solutions need some sort of power. The failure could also be as complex as two different interfaces not being able to talk to each other with sufficient efficiency to meet organizational goals. Failure mode and effect analysis for this situation should always include ratings of frequency and severity (Figure 26.1). A machine losing power should be low in frequency and very high in severity, whereas the

Failure Mode	Severity	Frequency	Mitigation for Failure
System not powering on	High	Low	Take steps ABCD

Figure 26.1

Severity	Description	Explanation
1	Least Often	1E-04
2	Not Often	2E-04
3	Sometimes	4E-04
4	Often	2E-02
5	Most Often	4E-02

Figure 26.2

machine not being able to integrate with different APIs would be a more frequent failure but lower in severity and able to be resolved very quickly (Figure 26.2).

Common failure methods for a logistics business might include the inability of a package or a human to get from point A to point B. This could be due to infrastructure constraints caused by a winter storm, an accident, or a literal roadblock of some kind. This means that the frequency of different failure modes will sometimes be geographically dependent and vary significantly from service region to service region.

You might need to execute failure modes and effect analysis for a very specific situation, germane to your specific geography, for example, traffic blocks and failure modes for getting an Uber from New Jersey to NYC in a given amount of time, using the Lincoln tunnel. The frequency of this failure mode would be high relative to a plane not being able to get across the ocean. But it is a less severe failure and, with some planning, a failure mode that could be largely avoided.

Lastly, we have the industrial use cases, which for our purposes in this book, we have assumed to be generally product focused. The industrial use cases identify when the desired density, the desired shape, or the desired weight of some kind of widget or industrial artifact is not achieved. When this happens, it would be a result of the failure of some kind of machinery. Ideally, this would be a low-frequency failure, but depending on the role of the specific product, the problem could be very high in severity, which is why quality assurance and other kinds of systems engineering tools come into play.

That is the failure mode and effect analysis. You rank items by severity, you rank items by frequency, and you identify the ways in which your selected system or selected perspective of a system will fail. This becomes the inverse of where most of the other tools focus, that is, on success and the happy path.

The failure mode and effect analysis and interfaces are two different ways to look at a system. For the most part, systems engineering relies on a system or a system's actions with external systems. There are very few tools in systems engineering, or any pedagogy for that matter, that focus on the inversion of action, which is lack of action or failure. Failure mode and effect analysis stands to provide a very concrete framework that a systems engineer can use to understand when their system will fail.

What I appreciate the most about the failure mode and effect analysis is that business owners, executives, and entrepreneurs do not always think about failure. The best ones embrace failure, they lean into it, but they do not often think about how they will fail. As an entrepreneur myself, I understand that failure confronts us on a daily basis. Some would say on an hourly or even second-by-second.

The failure mode and effect analysis tool is about focusing on the failure reasons that are within the control of the systems engineer and then planning for those both in terms of a severity level and an occurrence likelihood that can then help the engineer identify ways to mitigate those failures.

It is a basic ten-step process, which is to select items for the analysis. An example of that is the failure of a system because there is a global stock market collapse as opposed to the failure of a system because one of the engineers did not log in or submit a time card that day. One of those you can control, one of those you cannot as a systems engineer.

The ten steps for establishing a failure mode and effect analysis are:

1. Select items or functions for analysis.
2. Identify failure modes for each item.
3. Assess the potential impact of each failure mode.
4. Rate the severity of the potential impact.
5. Brainstorm possible causes for each failure mode.
6. Rate the likelihood of occurrence of each possible cause.
7. Suggest corrective actions for each possible cause.
8. Loop back to Step 4 (revisit the severity of the potential impact)
9. Assess the risk.
10. Prioritize the corrective actions.

The purpose of Step 1 is to really focus on what you can control and to identify the failure modes. If we are talking about a use case of being able to log in to the system, how could that fail? You could have the wrong password–username combination. You could have a broken query between the backend database and the front end. There are a bunch of different ways that a user logging on could fail.

Assess the potential impact of each failure mode. If the user cannot log in, you could say that impact is high, medium, or low based on your perspective. List all possible causes and rank the likelihood of occurrence. We have 10,000 logins a day, and only one of them has failed. That is one of 10,000 per day, which would be a relatively low occurrence. That could be a good example of an empirical way to determine the likelihood of the occurrence of a failure. Suggest corrective actions, that is very important, and then continue to loop back to Step 4 so that you can revisit the level of the potential impact on the failure mode.

The last step of the failure mode and effect analysis is to assess risk and prioritize corrective actions to resolve. Figure 26.3 shows how some systems engineers may choose to rank the severity

Severity	Description	Explanation
1	Minor	Variation does not affect performance to a level of 5 percent, repair takes less than 5 minutes, repair costs nothing, alteration has no effect on safety, and repair requires no skill level
2	Very low	Repair does not affect performance to 15 percent, repair takes between 5 and 10 minutes, repair costs nothing, repair has no effect on safety, and repair may require the skill level of a young adult
3	Low	Repair does not affect performance to 20 percent, repair takes between 10 and 15 minutes, repair costs nothing, repair does not affect safety, and repair may require the skill level of a young adult
4	Moderate	Repair does not affect performance to 30 percent, repair takes between 15 and 20 minutes, repair costs nothing, repair does not affect safety, and repair may require the skill level of an adult
5	High	Repair does not affect performance to 30 percent, repair takes between 15 and 20 minutes, repair costs some household items (e.g. glue), repair does not affect safety, and repair may require the skill level of an adult
6	Very High	Repair does not affect performance to 30 percent, repair takes between 15 and 20 minutes, repair costs some household items (e.g. glue), repair makes toy unsafe, and repair may require the skill level of an adult

Figure 26.3

description and explanation. An anecdotal example might be: "If, for failure reasons, the frequency can be seen in Figure A, B, C, or D, and it indicates the anticipated likelihood, the explanation, or the 1×10^{-4}." Of course, the systems engineer should come up with their own frequency based on empirical research.

Finally, the deliverable of the actual failure mode and effect analysis is the failure mode and what is being defined – the severity, the frequency, and then the mitigation of failure.

Chapter 27
Interface Matrix

This chapter is about interfaces and diagramming how interfaces connect to different systems or subsystems based on systems engineering principles. There are a variety of ways to represent interfaces in systems engineering pedagogy. I have selected a relatively simple method because it does allow you the reader to understand the philosophy of the tool. Understanding the philosophy can be just as useful as the tool itself. Simply put, to generate an interface matrix based on systems engineering tools, you line up the inputs on the rows, stacking them one on top of the other, and then you line up the outputs in the columns identifying what output is related to your system or subsystem.

You need to use some kind of marker. In Figure 27.1 I have used circles. One could use single circles, double circles, Harvey balls, or any kind of sphere that is shaded according to its level of severity, which would indicate a third dimension of thinking inside the two-dimensional representation I have presented you with in the figure. Here, I have simply included circles which serve to form as much as a check box.

To be clear, these examples are related to interface identification, not interface specification.

	Output 1	Output 2	Output 3	Output 4	Output 5
Input 1	●	●		●	
Input 2			●	●	
Input 3		●		●	
Input 4			●	●	●
Input 5	●	●	●	●	●

Figure 27.1

In this illustrative interface matrix, I am saying that input one interfaces with output one, and input one also interfaces with output two. You will see this as an example later on. Inputs can interface with all outputs. An input could be as simple as a website or some kind of account login, and an output could be as simple as a variety of different reports and dashboards.

The purpose of this tool is to make the systems engineer, entrepreneur, business owner, or executive understand how their system is connected. Their system is connected in more ways than one. It is connected in dynamic ways. That is where the Harvey ball or the moon – which is a filled circle – can be utilized to show that input one has a high or a medium level of connection with output one versus a low. When the moon is filled in, it shows one extreme, when it is completely empty, that is the other extreme. In this example, I have kept it simple, and used only full circles. But you can have partially filled or shaded moons.

If you consider an illustrative example such as a food-delivery system, you can consider the inputs being username and password, selected items from the food menu, tip amount, and delivery time, then the outputs from those inputs (Figure 27.2). An output from the username and password could be the log on screen and your order history. An output of the selected items could be the food menu, the order history, and the payment amount. An output of the

	Logon screen	Order History	Food Menu	Payment Amount
Username and Password	●	●		●
Selected Items		●	●	●
Tip Amount				●
Delivery Time	●			

Figure 27.2

tip amount is the payment amount, and an output of the delivery time could be on the log-on screen.

There is a very simple way to represent the user interface flow for a food-delivery system using the interface metrics.

Chapter 28
House of Quality

A House of Quality (HOQ) is a diagram used in engineering and product design to capture the voice of the customer and translate it into specific technical requirements for a product. The diagram is typically arranged in the shape of a "house," with a large rectangular shape at the top representing the customer needs and a series of smaller shapes below representing the technical requirements that must be met to satisfy those needs.

The HOQ diagram typically includes several key elements, including:

- Customer Requirements: A list of customer needs or wants, typically collected through market research or direct customer feedback.
- Technical Requirements: A list of specific technical requirements that must be met to satisfy each customer need. These requirements may include product features, performance metrics, or other specifications.
- Relationship Matrix: A matrix that shows the relationship between each customer need and each technical requirement,

indicating which technical requirements are most critical for satisfying each customer need.

- Competitive Assessment: An analysis of how well competing products or services currently meet each customer need and how well the proposed product or service is likely to compare.

By using a HOQ diagram, engineers and product designers can ensure that the design of their product is focused on meeting the specific needs and wants of their customers and that technical requirements are aligned with those needs. This can help improve customer satisfaction and ensure the success of the product in the marketplace.

In my opinion, the HOQ is not the most useful tool for systems engineering in business, but it is a very common tool for systems engineering. It would, hence, be a mistake to ignore it.

The house of quality can be a valuable graphic representation of how systems engineering principles can be overlaid with business operations or intangible assets.

The house of quality pictured in Figure 28.1 includes six main components:

1. Customer objectives,
2. Customer perceptions,
3. Engineering characteristics,
4. The impact of those characteristics,
5. Interrelationships,
6. End targets.

Customer objectives should be gathered based on market data or research – whatever executives are most concerned about – because our perceptions require the mapping of how competitors connect with customer requirements in a given marketplace.

Engineering characteristics identify how one measures customer perception. The impact of engineering characteristics shows how an increase or decrease in one characteristic affects customer attributes. For example, an increase in the number of accessories might make a toy more entertaining. The interrelationships show

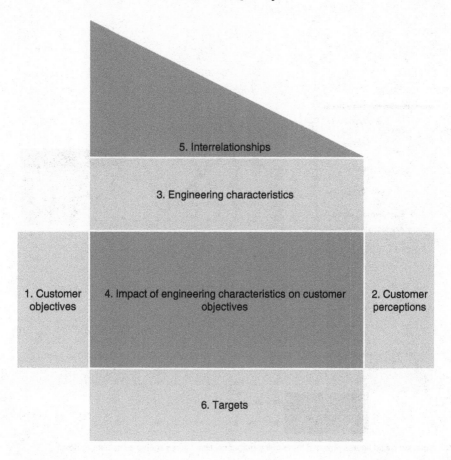

Figure 28.1

how an increase or decrease in one engineering characteristic affects another.

Finally, the targets identify benchmarks: units of measure and information about each engineering characteristic. You can see in Figure 28.2 an abstract representation of how these six variables come into play inside a single two-dimensional object.

In practice, a house of quality looks more like Figure 28.2.

It is a busy diagram, but it allows you in a single visual snapshot to understand how these six variables interact with each other and how changing one variable may change another.

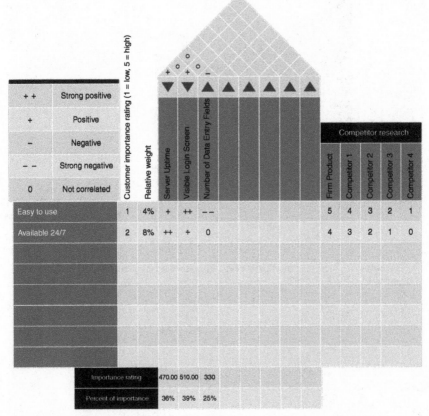

Note: simplified figures provided for learning purposes, final figures should contain more detail

Figure 28.2

For example, if you take the second row of this house of quality, you see we are talking about making the server online 24/7, which received the highest scale of relative importance of 8 percent. Notice the negative sign corresponding to the number of data entry fields, which indicates that there is a negative correlation between the server's availability and amount of data that the server is expected to upload and receive during use. Likewise, there is a positive correlation with the ease of use of the platform and the server uptime. You can see that in the first column, the customer wanted something easy to use more than something that was available 24/7.

As you increase the number of data fields entered, you reduce the platform's ease of use. Each of these variables ties straight back to customer perception. Remember the target benchmarks? Across the bottom, you have the objective measurements.

The house of quality can be cumbersome, and the added value of a house of quality chart is probably diminutive compared to the amount of work that it takes to create one. That said, no study of intangible assets would be complete without it. If you need to know how your intangible assets compare in the marketplace of other options, the house of quality becomes an irreplaceable systems engineering tool.

Chapter 29
Operations Description Template

A very important systems engineering tool is the operations description template, otherwise known as the ODT. The ODT is a template that shows how subsystems should behave from the viewpoint of various requirements and parties that interact with this system.

The purpose of developing the ODT is to make sure that subsystems complete necessary tasks so that the whole system can satisfy the needs of a customer. From this point of view, it is an important function of requirements validation.

Results from the ODT are inputs that create several tools for consolidation. Those tools are required for tracing matrices, functional interrelationship matrices, precedence matrices, functional flow diagrams, data transition matrices and diagrams, interface matrices, design and structure matrices, and others. The ODT itself is a base from which technical performance measures and bills of materials can be created.

The inputs to the ODT are use cases, behavioral diagram inputs, and substantive inputs. The ODT includes several use cases and functional requirements that relate to those. The ODT appears like a combined use case behavior diagram where systems are broken into different subsystems, and requirements and activities are recorded in the ODT.

To create an ODT, one must first review the systems of use cases, context, and requirements, then map the behaviors of these through the subsystems, identify messages, triggers, and interfaces, identify system state, set targets for behavior, extract functional requirements, and then trace drive requirements through originating requirements.

There are some important principles to follow when creating an ODP:

- It is better to have too many subsystem requirements included rather than too few.
- If the activity belongs solely to a single subsystem, there is no need to analyze this subsystem further.
- If the activity involves multiple subsystems, it should be analyzed into its constituent parts to see what each subsystem contributes to the overall functionality.

Most ODTs require many iterations to accurately reflect what the objectives of the systems are as mapped to the relevant inputs.

People often organize their ODTs in accordance with a sequence of use cases. For example, in the use case of a toy, think about a parent picking up batteries to put into a toy that gives it power. An ODT can read down a column from top to bottom, and interfaces can be shown in rows to articulate when a connection between systems is necessary.

The states that are identified within the ODT can be indicated on a recurring basis. The final column of an ODT gives additional information about the time that it may take to complete the chosen activity for a given subsystem.

The requirements stated in the ODT consist of previously developed functional requirements and newly developed functional requirements derived from the function of creating the ODT.

Figure 29.1 is an example of an ODT. We have an operator, which in this case is a parent picking up batteries, and we have different subsystems as defined through the context diagram, which

| Toy ODT | | | | | | | | | | | | | | |
| Controller | | | | | Cannon | | | | | External Components | | | | |
Operator	Controller Electronics	Controller Power	Controller Housing	Controller Attachment	Cannon Electronics	Target	Firing Mechanism	Cannon Power	Cannon Housing	ReCar	Car Controller	Projectile	State	Timing Component
Parent Picks Up the Batteries													Cannon (on, locked, fully loaded armed, unattached) Controller (off, not attached), system inactive	
Info Event, Batteries Picked Up		The controller shall be powered by batteries											Cannon (on, locked, fully loaded armed, unattached) Controller (off, not attached), system inactive	
		The controller shall be powered by three commercially available AA batteries												

Figure 29.1

are the controller, the canon, and the external components. On the right-hand side, you have the state that changes based on different actions made by the operator.

Interface rows are additional rows that exist in the ODT and show where some of the information can be coded into the system.

Not all interfaces engage with all columns, and these states change commonly based on the system that one is defining.

Chapter 30
Interface Tracking

The very last tool we are going to consider is interface tracking. Interfaces are the bedrock of a lot of systems engineering principles because, in this age of technology, all systems rely heavily on interfaces. One could even argue that a lot of the consumer product systems that exist in the economy today are merely an amalgamation or collection of APIs that are presented in an easy-to-use interface that helps the user accomplish a specific objective. This is essential to understand the role of the systems engineer and how they integrate into these frameworks.

The interface tracking methods most closely resemble the context diagram. You will remember from our exercises around the context diagram that the diagram evaluated external systems and how they are related to your systems. Interfaces are easiest to understand when you consider collections, inputs, and outputs of data between external systems and internal systems.

In the context diagram, we reviewed how an external system interacts with or manipulates the chosen system. In an interface tracking exercise, you determine the inputs and outputs that exist between a certain set of external systems and your chosen internal system.

Let us look at a few examples of interfaces. In the case of advertising, technology interfaces are extremely important. Interfaces provide a vast amount of data that can be placed into proprietary algorithms that help the advertising technology company drive consumer or user insight that an advertiser would be willing to pay for.

Interfaces inside a logistics company are essential because almost all logistics companies rely on a third party or several third parties to accomplish various functions.

In the case of Uber, the driver must have a connection with the Uber platform. The same goes with FedEx, which may be delivered to a third-party warehouse, where the user will need to have access to the building in order to deliver, file the forms, and so on.

Lastly, interfaces inside an industrial B2B example: If someone is selling an industrial product to multiple businesses, those interfaces would include customer feedback as well as different product information and might rely on sensors with Internet connectivity to collect information to complete closed-loop feedback for ongoing success.

Chapter 31
Conclusion to Tools

Hopefully, this overview of ten common systems engineering tools has helped you become more aware of some of the resources that are available to you as you evaluate intangible assets in your specific business use case.

Understanding and utilization of these tools will make you a more effective business owner, executive, engineer, or investor, capable of attaining a deeper understanding of the essential components of the system you are appraising or working within.

It is not my intention to present these ten tools as the final be-all and end-all of tools. Instead, it is my intention to help you understand that there are powerful frameworks and tools that already exist, which are highly developed within the pedagogy of systems engineering. I trust they will serve you well, allowing you to evaluate your own unique challenges with confidence.

Chapter 32

GQM, Decision Matrix, and Analytical Hierarchy for Ad-Tech Industries

To use the three tools of GQM decision matrices and analytical hierarchy for advertising technology, you need a very specific example. The tools are meant to elicit a set of solutions to solve a problem that might not be as easy to find using traditional business methods. For the example of advertising technology, I will use something that is very common inside digital advertising on streets, also known as digital out-of-home.

A very common problem with digital out-of-home advertising is the ability of a technology firm or solution to serve an ad to someone that is not necessarily in the digital environment. Juxtapose this with a person who is completely online. If you are on a website trying to buy European football jerseys, a logical ad to serve this person is something related to sports or something related to European or global football.

If a person is walking by a digital billboard, you may not have access to the exact state of mind that this human is in while they walk by; therefore, it is more difficult to serve advertisements that are completely relevant because the viewer or the consumer is offline.

Therefore, the context of placement in relation to user profiles becomes very difficult in a digital out-of-home scenario.

I want to focus – for this example of the GQM decision matrix and analytical hierarchy of advertising technology – on collecting relevant ad data of the human.

This becomes the goal. You have a digital billboard. It exists anywhere. It can be Times Square, Kansas, or China. But this digital billboard is there, it is connected to the Internet, and the firm behind it receives money to serve ads. The goal becomes "serve or place ads that are completely relevant to over 30 percent of the humans within viewing distance of a digital billboard." These questions are going to help us define how this goal is measured.

Questions include the size of the billboard and the geography around the billboard; a billboard in Times Square is very viewable from a certain angle. A billboard on a highway is visible from a different viewing angle.

Geography, size, and what is around the billboard. These criteria are very important if you have a digital billboard that is in a shopping world that has a very busy fast food chain is different from a billboard that is in the parking lot of a grocery store, which is again different from the Times Square example. Another question can be, what definition is available on the billboard in terms of the media type for the execution? Are you only going to be able to show very simple food and beverage advertisements or very rich and detailed content that requires a high level of definition? How will it display at night versus during the day? This is a common example of what people do not generally consider when placing out-of-home ads – what is the brightness level of the billboard? If we have a sign that is in direct sunlight, it will require a different brightness level than one on a cloudy night in a rainstorm.

These questions are very rudimentary, but they all point back to the way that you can create ideal metrics for this specific billboard. Each of these questions has ideal metrics. There are lumens; there are Nits (a Nit is a unit of measurement that equals one candela per square meter). A Nit measurement is how bright a screen appears to one human eye. The word "Nit" comes from the Latin word nitere, which means "to shine."

There are a number of different establishments around the billboard, how busy they are, how much foot traffic they have, what their annual revenue is, how many Google likes have, how many Google reviews they have, how many Instagram followers they have at that specific location, and so on. All of these can come together to create a series of ideal metrics for measurement of the goal, which is to serve this ad in a relevant way to more than 30 percent of the potential viewing public from where the ad is.

That is the goal, the questions that help you identify the goals, and then those are the ideal metrics that exist to support each of the goals.

Now you are coming to the decision matrix, and the decision matrix has to stem from the GQM. The decision matrix applies to only one goal of the GQM. So, if your goal is to provide an advertisement that is highly visible, that is the goal, and then the questions become the Nits of the screen, the placement of the screen, the geography of the screen, and so on. All of those can be weighed in accordance with one another, so you can rank them. If you have a billboard that sits atop a mountain that overlooks Hong Kong, you may indicate that the location of that screen matters at a number five, whereas the brightness of the unit matters less because it is so large.

If you have a screen that is very small, like at a gas station, brightness will matter more because the dimensions of the screen are small. You rate all of these in relation to each other, and each of these specific metrics or subgoals has its own product features and functionalities.

So, in the case of the brightness, it is not only the brightness, it is also the power supply; it is the definition in terms of pixels; it is the overall wattage. All of these play into the brightness. None of them impact the dimensions.

The weight. If you have a massive battery on top of a screen, then the weight becomes applicable to understand the brightness. All of them play together. These are sub-bullets. You are creating a hierarchy of examples. The goal at the top is to serve ads, and then you have the brightness in the next tier, and below that, you have Nits, dimensions, power supply, and so on, and you rank all of these in relation to each other as well. If you say that the brightness matters 60 percent, the weight matters 20 percent, the power matters 20 percent, you can multiply all three of these together to get the general weight of the brightness (Figure 32.1).

As long as your sub-bullets always add up to 100 percent, then your mid-level and eventually top-level will add up to 100 percent. This is how the analytical hierarchy process works. The process is nothing more than assigning relative weights based on the systems engineer, the business owner, or the executive's judgment.

The relative importance of each of the sub-points that came out of the decision matrix as related to the goal. What is so fascinating and powerful about this exercise is that now by using this simple question and probability tool that requires not even a calculator, you are relating the brightness of a screen to the number of fast-food chains that are nearby. That is how the hierarchy works.

What I love about the GQM decision matrix and analytical hierarchy process is that it is the only tool I have ever experienced in my professional career that is able to interrelate these components that you would think were never related. But they are related. When your objective is to serve a goal and the brightness matters, and the proximity to related business matters, and within the proximity, the

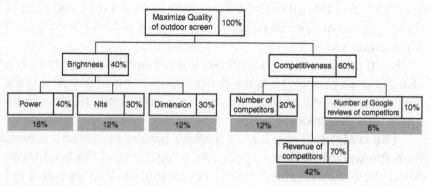

Figure 32.1

revenue matters twice as much as the number of establishments, you are able to mathematically relate those to the Nits of a screen, and there is no other tool inside of systems engineering or, in fact, the world that I have ever seen that allows any user to do this because this is reality as well. Reality is interrelating these variables that are one, two, three, and four steps removed. Computer programs can do it, too; they do it through very complex linear algebra, but those are not tools that are available to normal people, particularly with how business owners have to spend their time on a business class flight between San Francisco and New York City.

This is an example of the GQM decision matrix and analytical hierarchy within the context of advertising technology. To make sure that the point is driven home, I will take you through my examples on logistics and manufacturing in the following chapters .

Chapter 33

GQM, Decision Matrix, and Analytical Hierarchy for Logistics Industries

We reviewed the advertising technology example for the GQM decision matrix analytical hierarchy process. We now need to review the same set of tools for a logistics company like Uber, FedEx, or DHL.

Again, just like we did with the ad placement to over 30 percent of possible viewers being relevant, we now need to scope this thought exercise within the limits of a business objective.

In logistics with Uber or DHL, they have very, very different business models, and the easiest way to utilize the tools is to come up with something very simple that you understand so that you can judge the quality of the example I am giving you because you, in some ways, are an expert in this use case because you are a user. We will stick with the Uber example, and we will talk about a user who wants to get from point A to point B and wants to do it in the fastest, most efficient, comfortable, enjoyable, and cool way possible.

The top-level goal becomes getting from home to work via the most efficient route. The questions for this are going to be:

- How far away do they live?
- What are the traffic routes at this time?

- Can they pool, or can they use a direct route?
- Are they going on a reverse commute or a direct commute?
- Will they keep the driver waiting for long periods of time?
- Will the driver be able to pick up a return fair?

All of these are logical questions to ask that we will then need to assign metrics to as we get deeper into the example. For going from home to work, the metrics are very easy. The number of miles, time of day, some level of traffic monitoring, some level of frequency; is this five days a week, or is it one day a week? These are the sub-points that will go into the decision matrix that we will relate to a question or a set of metrics that one would not think were totally relevant. Now, remember, this is a person getting from point A to point B; metrics around that person's favorite color or type of beer are irrelevant. The focus then becomes: if the person is trying to get from point A to point B, and we have these metrics around time and distance, another metric would be the cost. They want to do it in an affordable way. Does the driver receive enough compensation to cover their fuel expenses? A sub-metric becomes, "Is the cost variable because of traffic or going in and out of New York City on a regular basis?" If you live in New York, you know that the traffic can be miserable, or it can be tolerable. All of these become the bottom layer of the hierarchy, and then we simply start assigning multiplication percentage points to them, making sure that they always add up to 100 percent. This is easy to do in Excel, as per Figure 33.1. So we have here a goal. The goal is to get from point A to point B in a cost-efficient manner. The sub-bullet here becomes available. Is the car regularly available to the consumer? That is one subgoal or a subdivision matrix. The other becomes cost and variability. The point of my bringing up variability here is that if it costs a dollar one day and then a thousand dollars the next day, I would not consider that efficient or cost-effective. The availability variable is the time that you must wait for a car. Another variable is the fastest route percent of the time – being able to confirm that you, in fact, took the fastest route at all times. Then, let us consider the ability of the driver to get a return route. These elements must add up to 100 percent.

You may say that availability is 80 percent of the game to you, and you do not care how much it costs. Cost is the remaining 20 percent. As long as all elements add up to 100 percent, the formula will work. We have to assume a closed-state system; otherwise, the math will not work.

Now imagine a scenario where these are flipped. Person B may say, "I don't care. I've got all the time in the world, but I don't have any money. So low cost is 80 percent important to me, but availability is only 20 percent."

For our example in Figure 33.1, we will go with this person who is more of a business rider. This person is busy, they are always late for work because they are stuck doing their hair. They do not have time to wait for a car. I am going to weigh that at 70 percent. I know that the route is relatively consistent for me, so I am going to weigh the fastest route at 20 percent. In regard to the driver's ability to get to return, the user does not care. That is the driver's problem, so we will call that 10 percent. All of these add up to one hundred. Then, it comes down to the cost and the standard deviation to perform the ride. Person B is on a very fixed budget, and the standard deviation to them is going to matter 60 percent, meaning that they cannot tolerate a standard deviation. Notice the way this is written is not that it is a high or low standard deviation, it is just the importance of the

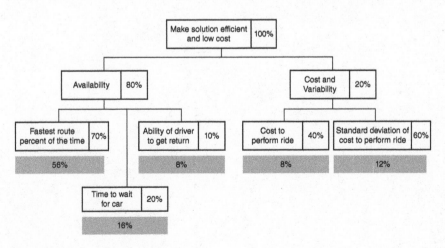

Figure 33.1

standard deviation within this system. Now, for this one, because it is the sub-point under the cost and variability example, multiply it by 20 percent, and all of a sudden, you have a sum of 100 percent across our beautiful example.

This means factually, mathematically, definitively, and absolutely, we can calculate that the cost of the ride matters one-seventh as much as the time this rider waits for the car, meaning that the time user-A spends waiting for a car is seven times more important than the cost.

You can say that absolutely, because of this tool that exists for our Uber example. There is no other tool in anything I have ever seen that allows a user to compare these variables that you would otherwise think were not comparable.

Chapter 34

GQM, Decision Matrix, and Analytical Hierarchy for Manufacturing Industries

We will now investigate the GQM, decision matrix, and analytical hierarchy tools for the manufacturing of a physical product or good in an industrial B2B use case.

We have used these tools up until now to focus on the general creation and ranking of criteria. The reason I have chosen to select these three different use cases for these tools is to show the variety that can exist for the ranking of these criteria.

This final example of the industrial use case, we will go through very quickly, and we will do that because the major difference between the examples we have already used is how a business makes decisions versus an advertiser – versus an individual.

So, let us start with the conceptual or strategic problem that an industrial product has to solve in order to adequately create a solution. Consider the durability of materials. Think about materials, coating, alloy treatment, industrial products that exist inside oil facilities and energy platforms, and. In this case, the goal would be to make a product more durable, let us say it is a screw. The subcategories under the durability of this will be the tensile strength, the weight, and perhaps the general size. Under those would be the

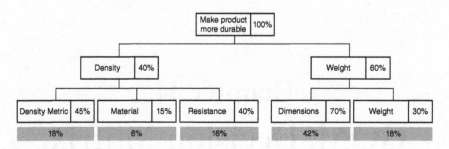

Figure 34.1

weight, density, and other criteria that are important to the out-come. Then, you can weigh all of those together to create a final score of what the customer may envision. As you can see in Figure 34.1, the goal is durability, and under durability, we are going to use density, and we are going to use weight, then under density, we have the density metric, the material, and the resistance to force. Under the weight banner, we simply have the weight and the dimensions. You assign relative scores to this. At the top level, density is worth 40 percent, and the weight matters 60 percent. Under that, dimensions might be 70 percent important, and the weight is the remaining 30 percent. Then, in density, you have the density metrics, the material, and the resistance, and you say that it is 45, 15, and 40 percent.

Voila, you have your GQM decision matrix and analytical hierarchy used to relate these tertiary metrics to the top-level metrics. It is that simple. I hope I have convinced you that these tools, used in tandem, allow the systems engineer, business owner, executive, or investor to relate metrics in ways that did not seem obvious prior to your discovery of these tools.

Part III

Geographical Application

Chapter 35
Introduction

We are now entering the third and final Part of this book. Until now, we have discussed intangible assets, what they mean, how they are measured, and how businesses, investors, and owners can use them to create a competitive advantage. We reviewed tools that are part of a systems engineering pedagogy. These tools can be used by executives, business owners, and investors to understand and identify intangible assets. We then reviewed several case studies that indicated how intangible assets – whose presence can be shown by the tools – impact capital markets and investment banking transactions.

We are going to take a pivot now to a seemingly new subject. The purpose of this Part is to be highly practical and almost entirely market driven. We are going to review macroeconomic perspectives on how these intangible assets can articulate themselves in a specific market.

I own a business that specializes in North America, Latin America, the Middle East, North Africa, and Southeast Asia; therefore, I have chosen these regions and these economies to analyze the intangible assets and the systems engineering principles and how they influence each other.

This Part is designed to be example driven. The utility of the intangible assets and the systems engineering tools we have talked about is not directly overlaid or directly applicable on a one-to-one basis with each of the following regional discussions. However, the reason I include this macroeconomic perspective here is to show the reader that intangible assets and systems engineering can apply to almost any business scenario and to consider these intangible assets and systems engineering principles in widely different contexts allows me to teach them with better specificity.

Broadly speaking, the best way to explain a concept to someone is always to talk about it within extremes, and these four markets that we are about to investigate are definitely different versions of extremes. Each of them has its own challenges, its own intangible assets, and will leverage systems engineering solutions differently, and each of them is relatively small except for the United States on a global scale.

I have specifically avoided China, India, and Africa, which are three of the more common regions to talk about when investigating these emerging economies. The reason I have left out these three markets is that they are so large that one book or a series of books would have to be dedicated to them in order to do them justice. This will become clear as we start looking at the size of markets like the Middle East and North Africa.

Macroeconomic Perspectives of Intangible Assets

The investment banking use case for this Part is largely going to be expansion driven. Investment banking or corporate development generally falls along the lines of capital placements and M&A transactions, but it can include joint ventures, strategic partnerships, licensing, and franchising.

All of these categories fall into the typical domain of corporate development, which does bleed into an overlap with investment banking. Therefore, the nomenclature that I will use in this section of the book is going to draw heavily on corporate development specifically as a broader definition of investment banking, whereas the prior sections of this book have been highly focused on investment banking transactions as they are typically defined by Wall Street.

Before we get into the details of each region, it is important to briefly define what I mean when I talk about joint ventures, strategic partnerships, franchising, and licensing. I will take them in reverse order.

Licensing is where you grant someone the right to use a solution, a product, software, a tool, or a process, and in exchange for that right to utilize, they pay the licensor money. It can be a payment in kind. It does not have to always be money. Payment in kind can be equity, land, or anything, but fundamentally, you are granting them the right to use something, and they pay tangible value back to you. It is often the case with licensing agreements that the licensor is receiving the benefit of the success of the licensee, meaning if you are licensing a technology to write books, then that technology would collect fees as it is utilized; as fees are collected for its utility by the licensee, they would then be shared with the licensor in some kind of a royalty or revenue share model.

Franchising is popular and very common in the United States with quick service restaurants such as McDonald's and Wendy's. Franchising is simply where the franchisor grants the franchisee the right to use its supply chain as well as its logo and brand to provide a product or service. The major difference between a franchisee and a license agreement is that a franchise agreement is broader. It includes licenses, but it also includes training. It may include a fee paid by the franchisee to the franchisor, whereas license agreements may not always include that.

Strategic partnerships are often used as a catch-all phrase to describe two companies working together to accomplish some kind of corporate development objective, which usually falls into a revenue growth opportunity.

A joint venture, as formally defined, is where two companies allocate intellectual property, know-how, and tangible or intangible assets into a new entity to create a third entity, that is the marriage between the two parent entities that created the joint venture. Personally, I use the term joint venture broadly to include strategic partnerships, franchising, and licensing, as well as to include distribution relationships and good old-fashioned sales.

So as I walk through the corporate expansion, corporate development, and investment banking ideas, I may use the term joint venture. When I say joint venture, I am specifically talking about two companies working together to accomplish a common corporate expansion objective, which is almost always related to revenue.

Corporate expansion can be non-revenue related for certain technology or data companies that are trying to gain regulatory approval or trying to gain some kind of data insight for advertising placement, but in almost all cases, the objective of two companies working together for expansion comes down to a revenue objective.

Now we will investigate the first of the four regions, namely, the Middle East and North Africa (MENA).

Chapter 36
The MENA Region

The Middle East and North Africa – MENA – is a remarkably unique, diverse, complex, and exciting part of the world. The region includes over fifteen nations, and some countries debate which nations belong in this classification. The purpose of this book is not to address those debates, but the most common inclusions for Middle East and North African countries are Saudi Arabia, the United Arab Emirates (UAE), Egypt, Morocco, Lebanon, and other surrounding nations (Figure 36.1).

There are over five hundred million people in the Middle East and North Africa. The region speaks seventeen dialects of Arabic and has a GDP less than half the size of the United States. What is very interesting about the Middle East and North Africa are the population impacts of the region. There are five hundred and fifty million people with a GDP significantly less than the United States, which has a population of approximately three hundred and fifty million people.

The MENA economies are driven by two major categories. The first and most well-known is energy, and this is largely a function of Saudi Arabia's presence in MENA, which includes the country

Figure 36.1

having the second largest reserves of oil in the whole world, second only to Venezuela. Saudi Arabia has a GDP of over USD eight hundred billion, rapidly growing to one trillion. The second category of trade or value of the economy of the MENA region is general trade or agriculture. There is a significant amount of food, livestock, and even airplane parts that are manufactured and transported between these regions, particularly on a per capita basis. It also produces a very high amount of industrial goods, regardless of the presence of energy and oil.

What makes me so intrigued by and interested in the MENA region is that you have a macroeconomy since it includes very different economies with very large players.

Egypt has over one hundred million people. The birth rate in many of the MENA countries is high and the proximity of these countries to each other is also very close, and it creates an integration of decision-making that does not exist in the United States. If you juxtapose the United States or North America with the Middle East and North Africa, the United States and Canada are largely extremely independent countries. Politics exists in all nations, but neither country is making decisions highly dependent on the other. When you look at the Middle East and North Africa, there are a few countries that have the majority of the power, and within those few countries, there are only two countries that carry almost all of the political, economic, financial, and general goodwill weight of the region. I am specifically talking about the Gulf Cooperation Council, which includes six countries, namely Bahrain, Oman, the United Arab Emirates, Saudi Arabia, Kuwait, and Qatar. Within those six countries, the UAE and Saudi Arabia are by far the most powerful and prevalent players on both a regional and a global scale.

Let us explore the power dynamic between a country like Egypt and a country like the UAE. Egypt has one hundred million people. The UAE has approximately eight million. This would be like Connecticut being able to boss around Canada. The UAE has a political, economic, and powerful sway over economies like Egypt. Connecticut has no power over Canada. Of course, this is an example mostly for the North American audience. The significant power dynamics that exist between these relatively small markets give the region pros and cons for how to do business there.

The pros include consolidated decision-making and a very clear hierarchy of value in the way countries are collecting money from specific regions. The cons include highly politicized decision-making and highly competitive processes to work with decision-makers.

Juxtapose these situations or examples with Connecticut and Canada, if a company was trying to leverage the diplomatic ties between Canada and Connecticut, there would be many different non-competitive ways to do this. Whereas, if a country specifically tried to develop or leverage the connections between the United Arab Emirates and Egypt, it is a very political and difficult dynamic, mainly because of the significant imbalance between the two markets.

The same imbalance exists among every single nation in the MENA region in different ways. The balance between Saudi Arabia and Egypt has different kinds of extremes. The balance between Saudi Arabia and the UAE has different extremes too. The relationships between Qatar and Saudi Arabia are generally highly political. Iran is a highly political and controversial character on a global scale, as are Iraq, Israel, and Lebanon. All of these countries that play in the same relatively small macroeconomic field have these severe political and economic relationships with each other that people in the Western Hemisphere or North America, namely business owners, investors, and executives, are not familiar with. That unfamiliarity creates gaps in knowledge that I will continue to describe within this chapter but may inspire further reading to delve deeper.

Before we move to the next section regarding the economic opportunities, to point back to intangible assets and systems engineering, it is important to understand the geography. The Middle East and North Africa are very strategically placed, and specifically, the Gulf Cooperation Council nations are very strategically placed between very large and powerful parts of the world. Russia has recently increased in notoriety based on its invasion of Ukraine. China has continued expansion – by what some perceived to be aggression – as it grows its influential borders outside of its natural borders. Of course, Europe and the United States all have to run through this region in some way or another.

In many ways, America or the West can benefit from the new Middle East. There is a bigger picture of the Middle East and North Africa that particularly came out of the 2020 COVID pandemic. The Gulf States and Israel provide the United States with an opportunity to participate in a global economy that is more centered on the Middle East post-pandemic. This can be a cornerstone of US policy. COVID and the pandemic threatened the mature economies of the United States and Europe and accelerated much of China's position on the world stage, at least politically. There are a few countries that were able to respond to the COVID pandemic as well as the Gulf Cooperation Council countries. This rise of the Middle East can be seen as a gateway between the United States and China, almost like a new Europe.

It is important that the United States not be afraid of celebrating diplomatic progress achieved with the GCC, particularly as it relates to the Abraham Accords. Increased diplomatic and commercial links in the region can reaffirm things like Mohammed bin Salman's Vision 2030 goals and even Mohammed bin Salman who was quoted in 2018 as saying the Middle East can become the new

Europe. Being the new Europe is something that motivates many GCC leaders, particularly as Washington, DC, and Beijing grow more postured with each other. The middle ground that Europe was during the cold war could be something that the Middle East plays starting now and into the future.

The Middle East knows this, and it can exploit this position. While the pandemic impacted many cities in the United States and Europe, the Middle East was able to evolve through it. It made investments in technology, infrastructure, and growth that positioned it to be this gateway between the East and the West. It is important that US politicians craft policies based on the reality that the Middle East has transitioned from a region largely defined by conflict to a region where the silk road is once again social, economic, and cultural.

The Middle East is rapidly progressing. Early in the COVID pandemic, three of the top seven countries with the most vaccinations were the UAE, Israel, and Bahrain. In addition to things like a successful vaccination campaign, the death rates in these countries were relatively low, and this was all done while key industries such as tourism remained open. This success is not a coincidence. It is a result of long-term thinking and planning.

Saudi Arabia has been working to triple its non-oil revenues and has invested billions into futuristic cities like Neom and The Line – a new standard in sustainable living. The UAE successfully completed a mission to Mars during the pandemic and recently announced plans to almost double Dubai's population in 2021. Gulf economies will benefit from low debt-to-GDP ratios, which will allow them to maintain growth while more developed and leveraged economies have struggled. The United States has a debt-to-GDP ratio of 100 percent, and in 2021, Saudi Arabia had a debt-to-GDP ratio of less than 30 percent. This means that these countries have spending power and that they can use their economies to take out cash and grow the nations.

China has seen the region as highly strategic and has built very deep links into it. It is important that US policymakers do the same. Many political and business leaders in the United States are driven by impulses that the region is conflict driven, and this is simply not the case. The United States and the world will inevitably be forced to accept that young, less populated nations have a lot to offer and can, in fact, bring a breath of fresh air into a lot of global economies.

Now, let us dive deeper into the role of the Middle East and North Africa, the GCC, the UAE, and the world. It has become undeniably clear that a business cannot be global without being in Dubai. Dubai is a crossroads between the East and the West, between cultures, between religions, and the government has positioned the city to become one of the centers of gravity in the world. You cannot be global without being in Dubai, the UAE.

I would like to take a minute right now and focus on the UAE specifically, to understand some of the dynamics of the economy and why the economy and the opportunities it presents are relevant from a corporate expansion perspective, particularly to investors, executives, and entrepreneurs. The GDP of the UAE is approximately four hundred billion dollars. Roughly 50 percent of that GDP is driven by oil revenues, and those oil revenues are largely collected by a small set of families. This is fine, and we have already talked about how this fact requires the country to diversify into new revenue streams for its macro economy, but it is important to understand that this also means that it is a relatively small market, particularly in comparison with economies like China, India, and the United States.

The UAE has a significant number of inhabitants that were not born and will most likely not die in the UAE. By some estimates, over 90 percent of the citizens or the inhabitants of the UAE are expatriates. This creates a very unique setting where the environment is highly professional and highly commerce driven, however, sometimes short-term thinking is a result. It is not a bad thing that so many people living in the UAE are not from the UAE and are not citizens of the UAE, but it is unique for such a wealthy country – on a per capita basis – to have so many of its producers from other parts of the world. This culminates in the Gulf Cooperation Council countries having several weaknesses as well as strengths.

The weaknesses of the GCC countries include the lack of diversity in their markets, relatively low ceilings for their capital markets and for companies to receive some kind of valuation or exit, and a relatively small economy. On the other hand, these economies can change very quickly; they can redefine and rebrand themselves to be relevant in a world driven by East and West politics – as we are seeing today with China, the United States, and Russia. They can also offer unique benefits to stakeholders and shareholders globally, such as they have done in the past by not having any income tax, although that was recently changed to include a VAT of 5 percent.

From an intangible asset and a systems engineering perspective, an analysis of the region is quite simple. The value of the UAE or the GCC, including Saudi Arabia, is the fact that it has access to so many large markets. The UAE can fly to India, and the United Kingdom, with relative ease compared to other markets like Japan, London, New York, and California. We have a lot of difficulties getting to these large markets, and despite the changes that the pandemic has created, in-person business is still a necessary part of a functioning enterprise. The UAE has access to very large populations. We have already discussed that the United States, with a population of three hundred and fifty million people – albeit with a large GDP – is much smaller than the MENA region, which includes a significant land mass with five hundred and fifty million people and has a higher growth rate from a population standpoint than the United States.

The intangible assets of the MENA region are opportunity, change, and growth. These three elements always come together to create capitalistic options for companies to make money in arbitrage or general business models. The utility of the systems engineer is applicable in understanding how to leverage these market dynamics of the UAE for profitability and success.

Chapter 37
The ASEAN Region

Now let us examine Southeast Asia. Broadly defined as the nations that include Singapore, Thailand, the Philippines, and Malaysia. In some cases, it includes Myanmar, Vietnam, Thailand, and so on, but the area of Southeast Asia that I want to focus on for this book is the ASEAN – the Association of Southeast Asian Nations.

The ASEAN region, much like the GCC, comprises six different economies. Those economies include Singapore, Thailand, the Philippines, Malaysia, Myanmar, Indonesia, Cambodia, and Brunei. Within the ASEAN region, there are ten member states. Within these ASEAN nations, I want to focus on Singapore, the Philippines, Indonesia, and Thailand. The reason I want to focus on these countries is that there are a lot of similarities to the Middle East. We will get into those presently, but the Middle East has a transitionary nature that has a lot of value, based on its proximity to other large markets, and Southeast Asia is the same. It has transitionary value because it is close to India and China. Politically, it is largely very neutral, but there are some strengths in the Southeast Asian economy that the Middle East could learn from.

The GDP of Indonesia is almost a trillion dollars, similar to the size of Saudi Arabia, but it is not as concentrated in energy and oil. Before we get into the specifics of the nations in the Southeast Asian subset that we want to talk about, it is important to look at Southeast Asia on a map (Figure 37.1).

Southeast Asia has relatively close proximity to India and China. Southeast Asia provides a very logical jumping-off point for executives, business owners, and investors to enter the Chinese markets. Southeast Asia has very limited corruption; it is very organized. The capital markets inside Singapore and other Southeast Asian nations are very sophisticated, even in the lower middle market side. Southeast Asian companies have regularly achieved relatively high capital market valuation and investment banking outcomes because of the sophistication and commitment of the government to these regions.

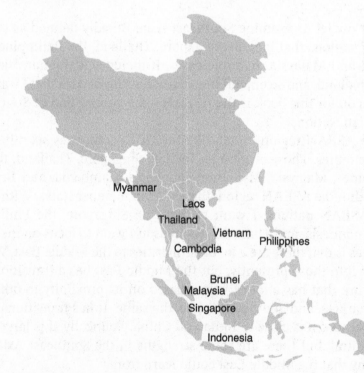

Figure 37.1

In summary, Southeast Asia gives American businesses or Middle Eastern and Latin American businesses three major opportunities: 1. size. They are large markets, particularly for relatively small companies that only do a few million dollars in revenue. 2. Southeast Asia is a transitionary market, very specifically focused on trade between countries because there is no single Southeast Asian country that is significantly large in itself. 3. Southeast Asia gives the capital-markets opportunity to businesses because it genuinely does have investors who are willing to put up capital and invest large sums of money. These three factors allow business owners to achieve lucrative outcomes.

*　*　*

I want to explain for a moment the value of the trade openness index, which is a metric used by global entities to determine how "open" an economy is to international trade. The trade openness index equation is very simple. It involves adding up a country's imports and exports and dividing it by the total GDP of the country. If a country has a GDP of 100 and it imports and exports 50, so its imports are 25 and its exports are 25, you will have 50 over 100, which is a 50 percent trade openness index. The trade openness index is just a simple metric that helps identify which economies are most active internationally.

Singapore has a trade openness index of over 300 percent, which means that if you add up the imports and the exports of the country, it is more than twice the GDP of the country, which is significant. The UAE, as we discussed in the previous section, is about 100 percent. When you compare the trade openness index to the United States, the United States has a trade openness index of only 26 percent. China has a trade openness of 36 percent. In 2019, Singapore was 300 percent, Saudi Arabia was 61 percent, Qatar was 86 percent, and the UAE was 160 percent. The trade openness index will generally be higher for countries that are smaller because if you are a small country, then the chance that you are importing or exporting more than you are essentially consuming domestically is higher. But it serves as a very strong indication of how the economy operates. Taking in just a very slight pivot from Southeast Asia, the African country of Djibouti has a trade openness index of 328 percent, and Djibouti is commonly seen as the next Singapore or UAE in the world.

Trade openness inside Indonesia is 37 percent, Malaysia is 123 percent, and the Philippines is 72 percent. The trade openness index is a very reliable metric to understand how active an economy is with external companies. What is interesting about the trade openness index of Southeast Asia is that when you look at several of the key countries within Southeast Asia, there are several that have very high trade openness indices. When you look at North America, Canada has a trade openness index of 65 percent, the United States is 26 percent, and Mexico is 78 percent. Compare this to Singapore, which is 300 percent and Indonesia at 37 percent. This approximately mirrors Canada and the United States. Malaysia is 123 percent, Cambodia is 123 percent, the Philippines is 72 percent and Thailand is 112 percent. This is significant because this indicates how countries work with and respond to those around them. The consolidation and concentration of high trade openness indices among Southeast Asian nations are unique. There are no other regions that have been developed where all of the countries have a very high trade openness index. Juxtapose Southeast Asia and Latin America, Brazil's trade openness index is 30 percent, Bolivia's is 56 percent, Argentina's is 32 percent, Chile's is 56 percent, Peru's is 47 percent, and Colombia's is 38 percent. They are all generally at around 50 percent or below.

Japan is 34 percent, Taiwan is 100 percent, China is 36 percent, India is 38 percent, Pakistan is 29 percent, and Kazakhstan is 63 percent. In 2019, Russia was at 50 percent. It is obviously far lower now.

There is a very high concentration of countries possessing a trade openness index inside Southeast Asia, and that fundamentally points to high trade activity, high frequency of trade, and an economy that is generally driven by more diversity – based on international exposure.

The takeaway for the business owner, executive, or investor is that these regions are open for business on a recurring basis, more so than anyone else inside of the regional analysis that we have conducted in this book.

Singapore will continue to experience incredible growth with the recent politics between Beijing and Hong Kong and Beijing's reduction of Hong Kong's independence. Asset managers and financial services companies have moved to Singapore because they feel there is potentially less risk than the more rigid regime of Beijing.

The presence of logistics, trade, and financial services – given the politics of Beijing, all moving into Southeast Asia – serve to make the region more relevant in the macroeconomic view of our study of intangible assets. The intangible asset of Southeast Asia can be defined very simply, it comes back to the trade openness index. When you have economies that are importing and exporting so much in relation to their own GDP, it creates opportunities for growth and arbitrage that do not exist in countries where everything is done domestically. This is particularly true for my firm, which specializes in cross-border transactions. The opportunity in Southeast Asia is very high for Jahani and Associates.

The three major things that make Southeast Asia attractive to investors, business owners, and entrepreneurs is that there truly is an opportunity for extreme growth in some of these countries.

When you look at the population of a country like Indonesia – over two hundred million people with a GDP of over USD one trillion, diversified, stable currency, and a large producer of palm oil – these kinds of countries can provide a plethora of opportunities that look very much like the scale of India. The growth in India from 2000 up until 2023 has been very extreme and generated a significant number of billionaires. It will be a lesser scale in countries like Indonesia or the Philippines, but the same opportunity is happening. These are also very tech-forward economies with very young populations. Extreme growth, diversity, and young populations make Southeast Asia very exciting.

The Southeast Asian region is political, not necessarily by choice but by absence. China – particularly in the wake of the Ukraine invasion by Russia and the COVID-19 pandemic that pivoted China against much of the world, but especially the United States – has made political decisions that have made Southeast Asia more relevant. Southeast Asia is a neutral territory between China and the United States, and it is my opinion that it will remain a neutral territory for the foreseeable future, but as China grows increasingly clustered against the United States, it is going to push commercial activity as well as US interests into the Southeast Asian region.

For an American business owner, the challenge with Southeast Asia is very simple, it is that it is very far away. The time zone difference is in excess of 12 hours for some Southeast Asian nations and for a business to thrive there you certainly have to travel there, and it is almost as foreign as you could possibly get. But from a cross-border perspective, ASEAN countries are fundamentally cross-border. They are very used to working on a cross-border framework with different cultures and different philosophies. The region could not be more primed for certain kinds of growth in e-commerce, logistics, and other technologies.

The most important opportunistic element of Southeast Asian economics is that it is very close to India and China, so it provides a very powerful and logical springboard for countries to enter India and China from that region.

Wrapping up our discussion of Southeast Asia, we have to consider the intangible assets that the region has as well as the tools that can apply to a systems engineer. The tools that I believe are most applicable to systems engineering within the Southeast Asian context come down to the customer. There are customer overlaps between Southeast Asia and the United States, namely in financial services and certain kinds of conglomerates. The style of doing business in most of Southeast Asia is more Western when compared to places like the Middle East, North Africa, and China, which can have very subtle nuances but nuances that are very pronounced throughout the region.

Doing business in Southeast Asia can feel more familiar to a Westerner, an American, or a Canadian doing business in parts of Saudi Arabia, Egypt, or Morocco. Therefore, the most applicable solutions that one can find when expecting or analyzing whether or not they should expand to Southeast Asia are customer-affinity processes, use cases, and behavioral diagrams. Entrepreneurs should have a very clear and defined thesis for why a Southeast Asian customer or a Southeast Asian vendor would be useful in the entrepreneur's overall strategy.

As discussed with intangible assets, the trade openness index is clearly an intangible asset of the region, and the presence of such intangible assets points towards specific industries such as cross-border e-commerce, logistics, tourism, hospitality, and travel, but less to domestic businesses such as regulated healthcare products.

Chapter 38
The Latin American Region

The Latin American, Hispanic, Central American, and South American worlds are difficult to analyze within the context of this book because the terminology is either too broad or too limited when trying to describe the Latin American nations that are relevant to the subject (Figure 38.1).

Before I get into the specifics of these regions, I want to take a moment to talk about some global dynamics that we see in cross-border investment banking. The terminology that I use to discuss cross-border investment banking is sometimes what some people might consider oversimplistic or insensitive. I just want to be clear that as I described these concepts and these tools, none of this terminology is meant to offend, belittle, or overlook any group anywhere in the world, I am simply communicating thoughts and ideas that utilize nomenclature that people do find in mainstream culture and academics even though that mainstream utilization may itself be insensitive at times. The examples of this will become clear as I unfold them.

There is a dynamic that we see in cross-border capital markets that is consistent with the terminology of the developed and

North America

Central America and The Caribbean

South America

Figure 38.1

neo-developed countries. I believe the terms emerging vs emerged, global south vs global north, developing vs developed, and so on are inaccurate terms to describe the dynamics we see in these varying economies. For this reason, I will use the terminology developed to refer to the more common, wealthy economies we are familiar with such as Europe, the United States, Canada, Australia, etc. I will use the term neo-developed to refer to the other countries that are the subject of this book such as the GCC and ASEAN countries. The fact is these economies are not emerging, they have already emerged; they are not developing, they have already developed. But they have emerged and developed differently, due to technology and globalization, thus my selection of the prefix "neo" to indicate they are "newly" developed. Neo-developed refers to a group of countries that are primarily located in the southern hemisphere and are generally considered to be less developed or developing in terms of their economy, infrastructure, and social indicators. This term is often used to replace older terms such as "Third World" or "developing countries," as it is more geographically descriptive and less derogatory.

Neo-developed is not a strictly defined region, but it usually encompasses countries in Latin America, Africa, and parts of Asia. These regions share some common historical, economic, and social experiences, such as colonialism, the struggle for independence, and continued challenges in terms of economic development, poverty, inequality, and political stability.

Some key characteristics of neo-developed countries include:

- Lower levels of economic development: Many countries in neo-developed regions have lower GDP per capita and human development index (HDI) scores compared to their counterparts in the developed countries.
- High levels of poverty and inequality: Neo-developed countries often struggle with widespread poverty, income inequality, and limited access to basic services such as healthcare, education, and sanitation.
- Limited infrastructure: Neo-developed countries may have less developed transportation, communication, and energy systems, which can hinder their economic growth and access to global markets.
- Political instability: Many neo-developed countries have faced historical and ongoing political turmoil, which can result in weak governance, corruption, and conflict.
- Dependence on primary industries: The economies of many neo-developed countries are largely based on the extraction and export of natural resources or agricultural products, making

them vulnerable to global market fluctuations and environmental degradation.

- Population growth: Neo-developed countries have higher population growth rates compared to the developed countries, which can put pressure on resources, infrastructure, and social systems.
- Environmental challenges: Neo-developed countries are often more vulnerable to climate change, natural disasters, and environmental degradation due to their geographic location and reliance on agriculture and other natural resources.

Despite these challenges, neo-developed countries are home to a diverse array of cultures, rich natural resources, and emerging markets with immense potential for growth and development. Many neo-developed countries are actively working to address these issues, with some achieving notable progress in recent years.

These neo-developed countries are often referred to as Latin America. They include Brazil, most of Africa, parts of Europe, parts of Turkey, and then parts of the Middle East. People can debate how much South Asia or East Asia fit this descriptor, but there certainly are parts of Southeast Asia that people would perceive as neo-developed. Canada, the United States, most of Europe, and for the most part, Russia and China are not referred to as neo-developed countries, and then, of course, you have Australia, which is not part of the economic designation of neo-developed.

There is a consistent framework that appears when you look at cross-border capital markets, specifically within the context of the developed countries and neo-developed designations. There are neo-developed markets that are close in proximity to developed country markets, and therefore, those neo-developed markets develop competitive advantages that the adjacent market will pay for, and that helps neo-developed countries markets grow and become competitive. I am referring to the GCC. The GCC, influences most of the MENA region. China influences a large part of Southeast Asia, and then, of course, you have the United States and Canada influencing almost the entirety of South and Central America, somewhat with the exclusion of Brazil.

Australia also impacts Southeast Asia, and Australia is very unique because it is isolated but has a global economic presence. The United Kingdom influences the whole world, even the United States. France has some connections to Lebanon, and so it goes on, but as you can see, this neo-developed and developed terminology is a very effective way to understand the kinds of competitive dynamics that appear in neo-developed countries' economies because of the opportunities that the developed countries' economies create. The reason I bring it up at this moment in the book is that it describes the relationship between South America, Central America, and the United States.

It is impossible to analyze Latin America, Central America, and South America without looking at the United States. The economy of the United States is twenty-two trillion dollars, larger than any South or Central American country by far. But before we get into the specifics of Latin America, let us explore the competitive advantages and strengths that each has and how those strengths articulate themselves differently in this neo-developed and developed dynamic.

In MENA, the neo-developed countries are largely the GCC, Saudi Arabia, Oman, Kuwait, Qatar, Bahrain, and the UAE. There are part of countries like Egypt that are neo-developed, though would be hard pressed to place Egypt and parts of MENA into the same category at the GCC. The power that Egypt has is its population which is more than twice as large as the entire GCC; almost one hundred and ten million people. Egypt is chaotic, with a volatile currency and a corrupt government, but the value that it brings to the table in this MENA dynamic is its size. When you think about food, payments, fintech, e-commerce, and logistics, Egypt is a very powerful player in the MENA space, and we have even seen this in the capital markets with companies like Swvl, which recently did a de-SPAC. Swvl is a ride-sharing program; we see a lot of opportunity in mobility as well.

One of the strengths of Southeast Asian countries is that they have relatively stable governments. Singapore is not largely corrupt. The countries of Southeast Asia produce a lot of global goods like palm oil, which is the number one export from Indonesia, and the Philippines. The Philippines produces a lot of labor and talent as well. The economies of Southeast Asia are more stable – on an anecdotal basis – than an economy like Egypt.

Now, we get into the specifics of Latin America. There are about six to ten Southeast Asian companies that are relevant. MENA has fifteen to twenty-five countries that are relevant. Latin America and Central America have many more. So, Latin America, Central America, and South America are more numerous, greater in size, and greater in disadvantages. It is also difficult to analyze the Latin American economy without talking about issues of corruption. The political status of most Latin American countries or South American countries is very unstable, volatile, and somewhat chaotic. The countries struggle with severe instances of corruption that have hamstrung and damaged the economies and the opportunities inside these regions for centuries and, unfortunately, will probably continue for at least a century or several decades to come.

The reality of corruption creates a weakness and an obstacle that changes the dynamic of the Latin American markets very significantly that does not exist in the MENA and Southeast Asian regions. This makes Latin America, South, and Central America very unique and very difficult to talk about with a broad brush, although we are still about to do it.

Mexico has a GDP of over one trillion dollars, over two hundred million people, and is right next to Texas. Mexico is, according to geographers, part of North America, and Mexico provides a wildly valuable location for agriculture, manufacturing, services, and other cost-center activities that the United States consumes. You are seeing a lot of integration between Hispanic and US media, individuals can just look at the number of songs on the top 50 hit list for Spotify and notice how many of them are in Spanish versus how many were in Spanish five or ten years ago.

As you go down the map and you move into Central America, similar dynamics to Mexico apply, although not as large. Here we find valuable outsourcing locations for most US companies and some recent political relationships with some Asian countries (specifically talking about Honduras' recognition of China, which is significant for the political relationship with Taiwan). These countries include Venezuela – which could be considered the Iran of the Western Hemisphere – Guyana, Suriname, Columbia, Peru, and so on. These countries have quite large economies that benefit from tourism, hospitality, agriculture, and so on, based on the climate as well as the cost of labor, and given the fact that Americans eat a lot, all of them are very focused on trade with the United States.

Brazil is the most unique and the most standout country in South Central Latin America. Brazil speaks Portuguese, and because of that, it is not considered a Hispanic country, which makes it difficult to pick the right term other than just rambling off South Central Latin America to describe the whole region in aggregate.

Brazil is a very significant economy. Their GDP is similar to the whole GCC: 1.6 trillion dollars, nearly twice the size of Saudi Arabia's. Brazil has a diversified economy but a somewhat weak currency. It has its own political instability and social problems, but it is certainly independent. For some of these Latin American and Central American countries, their existence is defined by their relationship with large buyers, which is mostly the United States. That is not the case with Brazil. Brazil is independent. In terms of land mass, it can stand up to the United States as a whole, which is unique for most countries since the United States is so big. It produces voluminous levels of financial services, labor, and products and services that are consumed globally, not just in Brazil.

So, Latin America and South and Central America are a very unique set of countries that play on the world stage differently than the other countries that we are analyzing in this book. The main weakness they have is corruption. In general, you can refer to it as weak or volatile political climates, but because of that volatility in the political climates, it creates a weakness that then impacts what they are able to do on a global stage. However, in terms of bringing in a cross-border advantage and a systems engineering approach,

Latin American countries can be extremely relevant and will continue to grow in relevance in the future.

Latin American countries have size, which is relevant, but more importantly, they have a cultural literacy with the United States. What I mean by cultural literacy is that the United States has been impacted by Latin America, South America, Central America, and Hispanic cultures so much that there is an ease of business for someone who is in Uruguay, Argentina, Mexico, or Honduras. This frictionless relationship does not really exist between someone in Indonesia and someone in China. Indonesia and China have highly distinct, ancient cultures that have been defined over thousands of years. The United States is a relatively new culture and is very much a melting pot, so for an entrepreneurial company, business owner, or an investor who has some kind of product or service that is relevant to the core market and cultural literacy that they are natively familiar with, they will be able to be competitive in the United States faster than someone in Indonesia would be with China. Similarly, in the Middle East, an Egyptian can be competitive in the UAE. There is quite a bit of cultural literacy between Egypt and the UAE due to being of the same religion. They both practice Islam, and the dialect of Arabic they speak is related. Egypt being a hundred-million population country, and the UAE being an eight-million population country, reverses the dynamic that someone from Honduras – which is an eight-million population country – would have with the United States, which is a three-hundred-and-fifty-million population country.

So, the dynamics between the United States and Latin America are very much in Latin America's favor. Whereas in other countries, the dynamics between GCC and MENA, those dynamics are in GCC's favor, the dynamics between East Asia and Southeast Asia, the dynamics are in East Asia's favor, in favor of China. This creates unique opportunities for Latin Americans to take advantage of in fields such as media, food and beverage, and particularly in e-commerce as Latin American familiarity in the United States continues to grow.

The systems engineering tools, much like we explored with Southeast Asia, come down to knowing your customer, which means entrepreneurs, investors, and business owners are able to understand use cases and behavioral diagrams. If you are a Latin American entrepreneur, investor, or executive reading this book thinking about which systems engineering principles might be most useful to evaluate your presence in the United States, use cases and behavioral diagrams are the most relevant tools because use cases are about how a customer utilizes a solution. The utility of how people consume Hispanic food, Hispanic media, or Hispanic fashion and textiles will be different from a company that is selling that solution in Latin America, Central America, and South America. Understanding these behavioral differences enables the Latin American company to service the United States more efficiently and effectively.

Chapter 39
Case Studies: Bringing It All Together

W e are now reaching the end of this book, and it is important to take a moment to remind ourselves of everything that we have analyzed as part of the exercises that we have shared. We have touched on three significantly large bodies of knowledge and three bodies of knowledge that are not often connected to one another in modern pedagogical, academic, or business literature.

First, we explored intangible assets as a function of capital markets and specifically how the intangible assets and the function of capital markets can be valued and can impact corporate value. We have talked about this within the scenarios of different investment banking outcomes, namely buy-side and sell-side M&A capital placement. Then, we explored the accounting definitions of intangible assets and how they can be used to articulate items that investors are familiar with. Second, we learned about systems engineering. As we defined early on our journey, systems engineering is a set of solutions that are largely used by aerospace engineers to build rockets, and yet we are effectively tying those tools to the questions, decisions, and problems that investors, business owners, and entrepreneurs have to solve.

205

This relationship between systems engineering and business is new, and it requires a mentality that is very data driven. Fundamentally, and at the end of the day, all engineering tools that I have presented to you require significant amounts of data. These are not strategy solutions. These are not qualitative behavioral analytic solutions. These are not qualitative human capital solutions. There are tools that apply to those qualitative fields that the user and reader can find in a plethora of different literature reviews. My objective in presenting these systems engineering tools is to make you think about how you can use these tools in data-rich environments.

Data-rich environments almost always rely on some kind of technology solution. It is very rare that you will find an application for systems engineering in business that does not apply to technology. Outside business, in engineering, mechanical, electrical, or almost any industry, it is very easy to apply systems engineering tools. But the fact of the matter is that business decision-makers, entrepreneurs, executives, and investors are often making very intuitive decisions for which they rarely have all of the data that they need.

Systems engineering can bridge this gap in select industries, which I have presented to you because of the rise in technology. Lastly, I have presented you with these macroeconomic use cases, which at first glance, may seem highly disparate and may seem like they are not related. The reason I have chosen to present these macroeconomic scenarios to you is that, as part of my investment banking practice, I have familiarity and expertise in these regions, additionally, because they are so different.

Latin America is so different from Southeast Asia, which is so different from MENA, which is so different from the US or North American countries. They are so different that it provides a really good academic basis for understanding intangibles.

It is difficult to do an analysis of intangibles and systems engineering when you are talking about a company that focuses on drone route lights, like flying drones across urban spaces with light routes, and a company that focuses on robots that operate within one's home to perform different household chores.

These two examples that I just anecdotally presented to you are so specific and really so interrelated that the value of the educational, academic, or philosophical discussion of systems engineering tied into capital markets and intangible assets tied into these macroeconomics, is lost.

That is why I have chosen to present to you these very large significant macroeconomic changes because it is very easy to understand how the Latin American market on a cross-border investment banking practice or objective is so different from Southeast Asia simply because Latin America is influenced by the United States and Southeast Asia is influenced by China.

Metaphorical extremes are always an effective way to educate and teach principles, and that is why I have chosen these three different examples. I wanted to make sure that you are familiar with why these examples have been chosen to deepen your understanding of how to utilize their value. It is important not to get overwhelmed by the details in these very large ideas that we are talking about. At the end of the day, what any business management or executive book you read, or any successful business leader will tell you is that you need data to make good decisions. It is your responsibility as the executive, the business owner, or the investor reading this book to collect that data. What I am trying to help you understand is that once you have the data, how can you start to use it to make meaningful insights, and how can you do it in a way that does not take you 100 hours? That is what systems engineering tools are designed to do. They are very good at it. That is what I want you to take away from this book.

Chapter 40
Case Studies: LATAM Expanding to the United States

I now want to share a series of client case studies based on companies that I have worked with as part of my investment banking and financial advisory business. These clients will represent different regions. They are issuers that came from the markets we have talked about such as North America, Southeast Asia, Latin America, and the Middle East-North Africa, and each of the examples and the case studies I offer will show you how clients were able to use systems engineering principles to impact and improve the performance of their business.

The first example I will share is a logistics company that was based in Latin America (LATAM), and this logistics company hired my firm to develop its distribution channels in North America. They sold consumer industrials. Things like home goods, supplies, and so on.

One of the challenges this company had in expanding its business relationships in North America was its weakness of being able to communicate its highly efficient and low-cost supply chain. It sourced most of its materials from Spanish-speaking countries, and it was able to sell those materials into the United States generating significant arbitrage. Most of its salespeople only spoke Spanish, and most of its executives were Hispanic men and women that had been brought up and developed mostly inside LATAM.

As the company worked through my firm to develop its capabilities and commercial presence in North America, it became very clear that there was a significant cultural gap between the buyers, who were largely of Anglo-Saxon descent, and the sellers, my clients, the Latinos. Cultural disparities in sales are not rare, and there are a plethora of consultants and advisors that help companies understand these cultural nuances and try to develop more competitive processes around them.

My firm decided to provide our client with a systems engineering approach, and what we did was conducted a customer affinity analysis, where we surveyed our customers. We collected over a thousand data points, and we grouped those customer responses into major categories. We focused on terminology and wording that made us believe certain cultural overlaps could exist, so instead of "cheap," we focused on clients that would use terms like "able to outsource" instead of "save money" or "inexpensive," and we focused on clients who use terms such as nearshore manufacturing.

As we went through this process with my firm's LATAM client, we were able to identify that, in fact, the set of customers we surveyed who were of a less Anglo-Saxon cultural persuasion was, in fact, looking for nearshore and were looking for outsourcing, because these were people who understood the benefits of working in other markets. With this data-generated realization, we pivoted the entire sales strategy of our LATAM seller, and we focused on targeting only companies that seemed to have some kind of Hispanic connection or affinity.

In summary, selling to people who reflected the results of the customer affinity analysis that understood the benefits of nearshoring or offshoring outweighed the potential risks or the potential changes in the business model.

It was a relatively simple exercise that our client did with us, and they were able to triple their sales over only a year and a half, but the benefits of using the customer affinity process for this client allowed us to point precisely to what the problem was – that the terminology and the utility of the less culturally integrated buyers were so different to what we were seeking that it became apparent we had to not only use a different set of nomenclature, but we also had to seek a different profile.

The client is very successful, and they have grown tremendously since we have been working with them. They are now selling across all major categories and all major cultures and added tens of millions of dollars of revenue.

This is what we focused on. This was a very simple tool that was quantitatively driven through the customer affinity analysis because we were able to group the comments, which allowed us to say 10 percent, 30 percent, or 80 percent of comments were in one category, where a different percentage was in a different category, which made it less valued. The customer affinity process is the only tool that allows you to execute this outcome so quantitatively. As I said at the beginning of this section, there are a seemingly infinite number of cultural tools people can use, but to literally add up the comments and then categorize those comments, is something that I have only seen in systems engineering and becomes a very definitive way to make these decisions. This could be equated, for ease of clarification, to using a word cloud or tag cloud to establish or visualize sentiment, but those are more pedestrian or domestic tools and are outside the field of the systems engineering tools we are exploring.

Chapter 41
Case Studies: MENA Medical Devices

T he next case study we will review is one from the Middle East and North Africa, namely the UAE. Jahani and Associates (J&A) work with a medical device company that provides chiropractic-recommended medical devices for lifestyle. This includes sleeping devices, home devices, and a variety of different solutions that individuals with back pain would need. This company was generating revenue in the United States – particularly through its partnerships with chiropractors – and it was interested in expanding to the Middle East and North Africa.

The company hired J&A for market expansion, and J&A was able to execute that mandate using both its investment banking acumen and its financial services skill set as well as its joint venture capabilities. The products that J&A's clients sold were specifically for people who had chiropractic issues for back pain. They hired J&A to conduct the market study of how the Middle East and North Africa could benefit the business as well as reach out to potential commercial partners who would buy the product.

The tools – the systems engineering tools that J&A used within this exercise – were very specific. We used the GQM decision matrix and analytical hierarchy processes. J&A created a series of goals that were specific to the company and used the GQM analysis to rank what paths for growth would be most applicable to the seller in this case. The initial thesis was that the company would be ripe for acquisition or some kind of capital markets objective.

After completing the GQM analysis and analyzing the potential returns by selling the asset or growing the asset inorganically versus just finding more sales partners to increase the natural sale of the business, the GQM decision matrix and analytical hierarchy process pointed – based on a board and shareholder consent – to growing the company through revenue in these markets. This was done through a variety of decisions, including the size of the addressable market, the desires and the strategic objectives of the shareholders, as well as the capabilities in the market for the product.

J&A was effectively able to broker a joint venture transaction with a large Abu Dhabi-based conglomerate and our client, the medical device company. At that time, it had represented an increase in sales of approximately 30 percent of the company's existing top-line revenue and became a platform through which the company was able to launch expansion efforts into India and China via this gateway UAE country. It also opened up Egypt and Saudi Arabia, which are more proximal to MENA and GCC.

The unique outcome of this exercise was that what J&A ended up doing was different from what the shareholders initially thought they wanted. This is often the case in lower middle market capital markets objectives. Companies often think they want to raise money or often think they want to buy an asset, to sell their assets, but in fact, they are not ready to do that, and the company can benefit from strategic partnerships, joint ventures, or direct sales rather than a capital markets activity. This is why I dedicated part of the book to corporate expansion, particularly within the concept of macroeconomics. Decisions simply become much clearer with the presence of data. Objectivity increases, and decision-making can be somewhat self-evident.

The macroeconomics of the MENA region is very different from the macroeconomics of the US region. The joint venture that J&A

specifically closed for our client, in this use case, was with a holding company that owned assets in hospitality, and they were able to buy the products sold by the medical device company in bulk, which represented a stepwise increase in revenue for our client. This would have been difficult to achieve for the same level of cost inside US markets. Labor and marketing costs in the United States are more expensive, and achieving the same level of distribution in the United States could not have been achieved with the single holding company to represent a 30 percent top-line revenue increase in the way that we accomplished this with our client in Abu Dhabi.

Chapter 42
Case Studies: ASEAN Aerospace Schema – Context Diagram

In this example, a Southeast Asian client hired Jahani and Associates to perform a valuation of their assets, specifically as they related to the space industry. This client was based in a Southeast Asian nation, and they developed a significant amount of technology that was proprietary to their system, and they believed it could have launched them into the next level of aerospace innovation.

J&A was hired both to build the company's desirability as an acquisition target and to indicate how the software that these engineers had developed was applicable to different teams within the aerospace framework. The deliverable that was developed heavily utilized context diagrams.

A context diagram, as has been explored deeply in parts of this book, is a way to show how an existing system interacts with external systems. The context diagram in and of itself is a relatively simple tool, but what it allowed J&A to do in the case of this aerospace example was indicate which external systems relied on data inputs from the core system that the engineers had developed.

By articulating this data need for these external systems, J&A was able to communicate the costs needed to develop the interfaces and provide direct data schemas, which included how they interacted and overlapped with requirements through both the external and internal system. Essentially, this allowed us to build a more fundable use case for a company which had a significant amount of technology, but had not developed any revenue traction. The intangible asset was the data schema and how it overlapped with the existing internal and external systems of our client. The impact was not that it made the valuation higher but that it gave the founders something to point to when articulating their value proposition, particularly given that the company was pre-revenue.

The deliverable was more than a context diagram in this example. The deliverable was largely about coding and detailed requirements that help to overlap the schemas between the external system and the internal system, but this data mapping and these schemas would have been very difficult to recreate, and – despite the fact that anyone who had them would have the same capability as our clients – they did give the client a significant competitive moat because it made it difficult for people to get access to data that they would need in order to form the schematics.

J&A's engagement with a particular Southeast Asian aerospace company ended in conversations regarding an acquisition. The client received an offer to be acquired by a publicly traded entity almost entirely for the intangible assets of the system and the firm that our client had built. The valuation methods for the schemas proved to be very useful in this example because – much like we have seen with the Direct TV acquisition and AT&T satellite orbitals being valuable – schemas and availability to integrate data systems were driving factors in this case.

The client worked with J&A to perform an additional valuation, not only of their schema assets but also other intangible assets that the company had developed, particularly with regard to how this specific acquirer would report the information to the public market.

This is a very important point to understand. There was a publicly traded acquirer that was in discussions with J&A's client, and in order to communicate the use case and the value for how the seller could improve the share price of the buyer, the seller continued to retain J&A to articulate those intangible asset values since the company had still not started to generate revenue.

To achieve this outcome, J&A utilized the exact tools outlined in this book. We reviewed the public filings of how this particular buyer had communicated the value of its own business to public market investors. We conducted a sensitivity analysis to determine when certain value statements influenced stock price positively, when they had no effect, and when they influenced stock price negatively; then we continued to value the specific assets of the seller and how those assets would create a tangible, numerical, and factual basis for the buyer to make statements that we deemed to be most likely received in a positive light by the market based on our capital markets analysis.

It was a combination of capital markets and public equities research combined with hardcore systems engineering – as it related to aerospace design – that then utilized intangible asset analysis, with specific regard to schemas such as orbital placement, trajectories into space, and general payload capacities. All three of these components came together to create a very valuable outcome for this aerospace client case study.

You can see from this example – and it is an important takeaway from this whole book – that the tools I am providing are not a way to artificially inflate value. They are a way to communicate the value that genuinely exists, as in the case of the schema. I often encounter entrepreneurs who are delusional and who – instead of creating value – focus only on the narrative and only on the marketing slogans rather than creating real intangible value that can be communicated to a public entity.

That is the most important thing for the investor, the business owner, or the executive to take away from the utilization of these tools. They are not a replacement for value. They are simply a way to measure and communicate the value that is not on a financial statement, which is the most fundamental and simple way to understand what an intangible asset is. It is the value that you cannot readily find on the income statement, the cash flow statement, or the balance sheet.

Chapter 43
Case Studies: In Conclusion

Within these case studies, I have presented examples of three different clients in three different markets with three different sets of systems engineering tools. The engineering tools allowed Jahani and Associates to provide management consulting value to its clients that bridge the gap between intangible assets and capital markets.

You must remember that the fundamental premise of the utility behind all the systems engineering tools is that they provide a framework for analyzing intangibles that no other pedagogy does. At least none that I have ever experienced. Tangible assets are very clearly understood through financial statements. Intangible assets almost always require an additional set of frameworks, and I believe that systems engineering can apply to and provide solutions for the vast percentage of business questions and reporting standardization that many intangible asset-heavy companies need.

Chapter 44
Summary

I learned early in my communications training to "tell 'em what you are going to tell 'em, tell 'em, then tell 'em what you told 'em."

I promised to arm you with tools that would collect the required data, identify hidden secrets in that data, show you how to exploit the uncovered genius in your data, and how to execute actions on the intelligence acquired.

I delivered complete how-to guides on each of the tools, when to use them, and who would be interested in what they produced.

We also explored whom to target with your newfound valuations due to unexplored intangible assets and the best ways to monetize or otherwise leverage these assets.

Then, like global adventurers, we explored geographic regions for potential expansion of the empire you will hopefully now grow to heights heretofore unimagined.

I believe wholeheartedly that with what I have shared in these pages, you will be better armed than the majority of entrepreneurs, business owners, and investors to supercharge what you have built or intend to build in your innovation-focused, profit-driven, or general entrepreneurial life.

In your future journey, when potential partners, investors, lenders, or acquirers ask to see your balance sheet and profit and loss but not explore your intangible assets strategies, you will know exactly who you are dealing with and how to amaze and bewilder them with your unexplored potential and knowledge of the business growth landscape at the very highest level.

I make myself and my company available to you to explore possibilities together and scale new peaks of knowledge and potential as our worlds and our world continue to break new ground in the unlimited realm of what is possible. Lastly, I will leave you with this thought composed by Dr. Serhan Ili.

"Imagine a 1000-gram iron bar. Its raw value is around $100. If you decide to make horseshoes, its value may increase to $250. If, instead, you decided to make sewing needles, the value might increase to about $70,000. If you decided to produce watch springs and gears, the value could increase, somewhat unexpectedly to about $6 million. However still, if you decided to manufacture precision laser components out of it like ones used in lithography, it could convert that humble $100 iron bar into $15 million.

Your value is not just what you are made of but in what ways you can make the best of who or what you are – your intangible value."

The End

Bibliography

Adam, M.B. and Donelson, A. (2022). Trust is the engine of change: a conceptual model for trust building in health systems. *Systems Research and Behavioral Science* 39(1): 116–127. https://doi.org/10.1002/sres.2766

Amico, A. (2016). From regulation to enforcement of corporate governance in the Middle East and North Africa. In *The Handbook of Board Governance: A Comprehensive Guide for Public, Private and Not-for-Profit Members*, ed. R. Leblanc, pp. 776–802. Hoboken, NJ: John Wiley & Sons. https://doi.org/10.1002/9781119245445.ch39

Anand, S. (2007). Corporate governance in emerging markets: Asia and Latin America. In *Essentials of Corporate Governance*, pp. 155–162. Hoboken, NJ: John Wiley & Sons. https://doi.org/10.1002/9781118385210.ch11

Andre, S. and Sanghvi, S. (2005). Feedstock price volatility and how to deal with it. In *Value Creation: Strategies for the Chemical Industry*, ed. F. Budde, U.-H. Felcht, and H. Frankemölle, pp. 201–214. Weinheim: Wiley-VCH. https://doi.org/10.1002/9783527612246.ch16

Andre, S., Sanghvi, S., and Röthel, T. (2005). An approach to determining the long-term attractiveness of commodity chemical businesses. In *Value Creation: Strategies for the Chemical Industry*, ed. F. Budde, U.-H. Felcht, and H. Frankemölle, pp. 63–77. Weinheim: Wiley-VCH. https://doi.org/10.1002/9783527612246.ch6

Apte, S., Lele, A., and Choudhari, A. (2022). COVID-19 pandemic influence on organizational knowledge management systems and practices: insights from an Indian engineering services organization. *Knowledge and Process Management* Special issue. https://doi.org/10.1002/kpm.1711.

Balkaran, L. (2012). Bahamas. In *Directory of Global Professional Accounting and Business Certifications*, pp. 15–22. Hoboken, NJ: John Wiley & Sons. https://doi.org/10.1002/9781119198628.ch2

Balkaran, L. (2012). Umbrella organizations. In *Directory of Global Professional Accounting and Business Certifications*, pp. 227–235. Hoboken, NJ: John Wiley & Sons. https://doi.org/10.1002/9781119198628.app1

Barrows, E. and Neely, A. (2012). Understanding turbulence. In *Managing Performance in Turbulent Times: Analytics and Insight*, pp. 1–20. Hoboken, NJ: John Wiley & Sons. https://doi.org/10.1002/9781119202547.ch1

Biegelman, M.T. and Biegelman, D.R. (2008). The international landscape of compliance. In *Building a World-Class Compliance Program: Best Practices and Strategies for Success*, pp. 107–130. Hoboken, NJ: John Wiley & Sons. https://doi.org/10.1002/9781118268193.ch6

Biegelman, M.T. and Biegelman, D.R. (2010). Worldwide hotspots for corruption: UK, Russia, Africa, the Middle East, and Latin America. In *Foreign Corrupt Practices Act Compliance Guidebook: Protecting Your Organization from Bribery and Corruption*, pp. 143–170. Hoboken, NJ: John Wiley & Sons. https://doi.org/10.1002/9781118268292.ch6

Budde, F., Felcht, U.-H., and Frankemölle, H. (2005). Today's challenges and strategic choices. In *Value Creation: Strategies for the Chemical Industry*, ed. F. Budde, U.-H. Felcht, and H. Frankemölle, pp. 53–61. Weinheim: Wiley-VCH. https://doi.org/10.1002/9783527612246.ch5

Camman, C., Fiore, C., Livolsi, L., and Quessa, P. (2022). Other titles from iSTE in systems and industrial engineering – robotics. In *Supply Chain Management and Business Performance: The VASC Model*, pp. G1–G9. London: iSTE; Hoboken, NJ: John Wiley & Sons. https://doi.org/10.1002/9781119427407.oth2

Chakravarty, V. and Ghee, C.S. (2012). Asia rewrites the M&A rules. In *Asian Mergers and Acquisitions: Riding the Wave*, pp. 1–9. Singapore: John Wiley & Sons. https://doi.org/10.1002/9781119199120.ch1

Chakravarty, V. and Ghee, C.S. (2012). Asian companies are poised to triumph in the merger endgame. In *Asian Mergers and Acquisitions: Riding the Wave*, pp. 11–29. Singapore: John Wiley & Sons. https://doi.org/10.1002/9781119199120.ch2

Chakravarty, V. and Ghee, C.S. (2012). Index. In *Asian Mergers and Acquisitions: Riding the Wave*, pp. 197–206. Singapore: John Wiley & Sons. https://doi.org/10.1002/9781119199120.index

Chakravarty, V. and Ghee, C.S. (2012). Notes. In *Asian Mergers and Acquisitions: Riding the Wave*, pp. 189–194. Singapore: John Wiley & Sons. https://doi.org/10.1002/9781119199120.notes

Chakravarty, V. and Ghee, C.S. (2012). The rise and rise of cross-border M&A. In *Asian Mergers and Acquisitions: Riding the Wave*, pp. 31–52. Singapore: John Wiley & Sons. https://doi.org/10.1002/9781119199120.ch3

Chakravarty, V. and Ghee, C.S. (2012). Transforming government-linked companies through mergers and acquisitions. In *Asian Mergers and Acquisitions: Riding the Wave*, pp. 53–68. Singapore: John Wiley & Sons. https://doi.org/10.1002/9781119199120.ch4

de Graaf, R.S. and Loonen, M.L.A. (2018). Exploring team effectiveness in systems engineering construction projects: explanations why some SE teams are more effective than others. *Systems Research and Behavioral Science* 35(6): 687–702. https://doi.org/10.1002/sres.2512

de Mahieu, C., Günther, C., and. Riese, J. (2005). Middle East: opportunities and challenges from the rapid emergence of a global petrochemical hub. In *Value Creation: Strategies for the Chemical Industry*, ed. F. Budde, U.-H. Felcht, and H. Frankemölle, pp. 79–93. Weinheim: Wiley-VCH. https://doi.org/10.1002/9783527612246.ch7

Dodel, K. (2012). Valuing private companies — the PCD. In *Private Companies: Calculating Value and Estimating Discounts in the New Market Environment*, pp. 67–172. Chichester: John Wiley & Sons. https://doi.org/10.1002/9781119960508.ch3

Dugan, K.E., Mosyjowski, E.A., Daly, S.R., and Lattuca, L.R. (2022). Systems thinking assessments in engineering: a systematic literature review. *Systems Research and Behavioral Science* 39(4): 840–866. https://doi.org/10.1002/sres.2808

Fabiano, P. (2008). Compliance in Mexico: trends, best practices, and challenges. In *Governance, Risk, and Compliance Handbook*, ed. A. Tarantino, pp. 839–854. Hoboken, NJ: John Wiley & Sons. https://doi.org/10.1002/9781118269213.ch60

Fabiano, P. (2008). Global compliance programs in Latinamerica: major challenges and lessons learned. In *Governance, Risk, and Compliance Handbook*, ed. A. Tarantino, pp. 645–660. Hoboken, NJ: John Wiley & Sons. https://doi.org/10.1002/9781118269213.ch47

Feinschreiber, R. and Kent, M. (2012). China–Taiwan trade. In *Transfer Pricing Handbook: Guidance on OECD Regulations*, pp. 345–356. Hoboken, NJ: John Wiley & Sons. https://doi.org/10.1002/9781119203650.ch24

Fontaine, R. and Ahmad, K. (2012). Case studies. In *Strategic Management from an Islamic Perspective: Text and Cases*, pp. 151–154. Singapore: John Wiley & Sons. https://doi.org/10.1002/9781118646007.part2

Frank, M., Sadeh, A., and Ashkenasi, S. (2011). The relationship among systems engineers' capacity for engineering systems thinking, project types, and project success. *Project Management Journal* 42(5): 31–41. https://doi.org/10.1002/pmj.20252.

Frankel, M.E.S. and Forman, L.H. (2017). Deal process. In *Mergers and Acquisitions Basics: The Key Steps of Acquisitions, Divestitures, and Investments*, 2e, pp. 107–133. Hoboken, NJ: John Wiley & Sons. https://doi.org/10.1002/9781119380726.ch6

Gregory, M. (2015). Manufacturing systems engineering. In *Wiley Encyclopedia of Management*, 3e, Hoboken, NJ: John Wiley & Sons. https://doi.org/10.1002/9781118785317.weom100115

Greve, H., Rowley, T., and Shiplov, A. (2012). The first-degree perspective: strengthening the foundation of network advantage. In *Network Advantage: How to Unlock Value from Your Alliances and Partnerships*, pp. 73–90. San Francisco: Jossey-Bass. https://doi.org/10.1002/9781118561393.ch3

Gulapalli, E.K., Morand, C.L., and Wolski, G.E. (2017). Merger and acquisition transaction disputes. In *Litigation Services Handbook: The Role of the Financial Expert*, 6e, ed. R.L. Weil, D.G. Lentz, and E.A. Evans, pp. 24.1–24.35. Hoboken, NJ: John Wiley & Sons. https://doi.org/10.1002/9781119363194.ch24

Gutmann, A. (2012). Introduction to investment banking. In *How to Be an Investment Banker: Recruiting, Interviewing, and Landing the Job*, pp. 1–53. Hoboken, NJ: John Wiley & Sons. https://doi.org/10.1002/9781119204992.ch1

Handfield, R. and Linton, T. (2017). The future of supply chains. In *The LIVING Supply Chain: The Evolving Imperative of Operating in Real Time*, pp. 175–195. Hoboken, NJ: John Wiley & Sons. https://doi.org/10.1002/9781119308027.ch8

Hansen, J.M., Morand, C.L., and Wolski, G.E. (2012). Merger and acquisition transaction disputes. In *Litigation Services Handbook: The Role of the Financial Expert*, 5e, ed. R.L. Weil, D.G. Lentz, and D.P. Hoffman, pp. 21.1–21.30. Hoboken, NJ: John Wiley & Sons. https://doi.org/10.1002/9781119204794.ch21

Holliday, M. (2017). Cultivating excellence: moving beyond engineering thinking to living-systems thinking. *Performance Improvement* 56(4): 27–30. https://doi.org/10.1002/pfi.21711

Holm, M. (2016). Cross-border M&A: region- and country-specific trends and deal planning tips. In *Cross-Border Mergers and Acquisitions*, ed. S.C. Whitaker, pp. 111–123. Hoboken, NJ: John Wiley & Sons. https://doi.org/10.1002/9781119268451.ch5

Jeddi, H. and Doostparast, M. (2016). Optimal redundancy allocation problems in engineering systems with dependent component lifetimes. *Applied Stochastic Models in Business and Industry* 32(2): 199–208. https://doi .org/10.1002/asmb.2144

Khayal, I.S. and McGovern, M.P. (2021). Implementing patient-centred behavioural health integration into primary care using model-based systems engineering. *Systems Research and Behavioral Science* 38(2) 246–256. https://doi.org/10.1002/sres.2727

Knudson, C. (2012). Interrupted production. In *Bribery and Corruption Casebook: The View from Under the Table*, ed. J.T. Wells and L. Hymes, pp. 65–73. Hoboken, NJ: John Wiley & Sons. https://doi.org/10.1002 /9781119204718.ch7

Krumsiek, B. and Kruvant, M.C. (2012). The business of empowering women: innovative strategies for promoting social change. In *The ICCA Handbook on Corporate Social Responsibility*, ed. J. Hennigfeld, P. Pohl, and N. Tolhurst, pp. 209–222. Chichester: John Wiley & Sons. https://doi .org/10.1002/9781119202110.ch13

Kwansah-Aidoo, K. and Saleh, I. (2017). A battle for hearts and minds: dealing with Syria's intractable humanitarian catastrophe. In *Culture and Crisis Communication: Cases from Nonwestern Perspectives*, ed. A.M. George and K. Kwansah-Aidoo, pp. 241–269. Piscataway, NJ: IEEE Press. https:// doi.org/10.1002/9781119081708.ch14

Lev, B. (1999). R&D and capital markets. *Journal of Applied Corporate Finance* 11(4): 21–35. https://doi.org/10.1111/j.1745-6622.1999 .tb00511.x

Lev, B. (2007). Discussion of "information asymmetry and cross-sectional variation in insider trading." *Contemporary Accounting Research* 24(11): 233–234. https://doi.org/10.1111/j.1911-3846.2007.tb00103.x

Lev, B. (2017). Evaluating sustainable competitive advantage. *Journal of Applied Corporate Finance* 29(2): 70–75. https://doi.org/10.1111 /jacf.12234

Lev, B. and Gu, F. (2016). Accounting: facts or fiction. In *The End of Accounting and the Path Forward for Investors and Managers*, pp. 94–103. Hoboken, NJ: John Wiley & Sons. https://doi.org/10.1002/9781119270041.ch9

Lev, B. and Gu, F. (2016). Author index. In *The End of Accounting and the Path Forward for Investors and Managers*, pp. 243–245. Hoboken, NJ: John Wiley & Sons. https://doi.org/10.1002/9781119270041.indauth

Lev, B. and Gu, F. (2016). Corporate reporting then and now: a century of "progress." In *The End of Accounting and the Path Forward for Investors and Managers*, pp. 1–13. Hoboken, NJ: John Wiley & Sons. https://doi .org/10.1002/9781119270041.ch1

Lev, B. and Gu, F. (2016). Epilogue. In *The End of Accounting and the Path Forward for Investors and Managers*, p. 241. Hoboken, NJ: John Wiley & Sons. https://doi.org/10.1002/9781119270041.epil

Lev, B. and Gu, F. (2016). Finally, for the still unconvinced. In *The End of Accounting and the Path Forward for Investors and Managers*, pp. 61–66. Hoboken, NJ: John Wiley & Sons. https://doi.org/10.1002/9781119270041.ch6

Lev, B. and Gu, F. (2016). Implementation. In *The End of Accounting and the Path Forward for Investors and Managers*, pp. 199–212. Hoboken, NJ: John Wiley & Sons. https://doi.org/10.1002/9781119270041.ch16

Lev, B. and Gu, F. (2016). Investors' operating instructions. In *The End of Accounting and the Path Forward for Investors and Managers*, pp. 230–239. Hoboken, NJ: John Wiley & Sons. https://doi.org/10.1002/9781119270041.ch18

Lev, B. and Gu, F. (2016). Matter of fact. In *The End of Accounting and the Path Forward for Investors and Managers*, p. 27. Hoboken, NJ: John Wiley & Sons. https://doi.org/10.1002/9781119270041.part1

Lev, B. and Gu, F. (2016). Practical matters. In *The End of Accounting and the Path Forward for Investors and Managers*, p. 197. Hoboken, NJ: John Wiley & Sons. https://doi.org/10.1002/9781119270041.part4

Lev, B. and Gu, F. (2016). Sins of omission and commission. In *The End of Accounting and the Path Forward for Investors and Managers*, pp. 104–111. Hoboken, NJ: John Wiley & Sons. https://doi.org/10.1002/9781119270041.ch10

Lev, B. and Gu, F. (2016). So, what to do with accounting? In *The End of Accounting and the Path Forward for Investors and Managers*, pp. 213–229. Hoboken, Nj: John Wiley & Sons. https://onlinelibrary.wiley.com/doi/10.1002/9781119270041.ch17

Lev, B. and Gu, F. (2016). So, what's to be done? In *The End of Accounting and the Path Forward for Investors and Managers*, pp. 113–118. Hoboken, NJ: John Wiley & Sons. https://doi.org/10.1002/9781119270041.part3

Lev, B. and Gu, F. (2016). Strategic resources & consequences report: Case No. I: Media and entertainment. In *The End of Accounting and the Path Forward for Investors and Managers*, pp. 133–145. Hoboken, NJ: John Wiley & Sons. https://doi.org/10.1002/9781119270041.ch12

Lev, B. and Gu, F. (2016). The widening chasm between financial information and stock prices. In *The End of Accounting and the Path Forward for Investors and Managers*, pp. 29–40. Hoboken, NJ: John Wiley & Sons. https://doi.org/10.1002/9781119270041.ch3

Lev, B. and Gu, F. (2016). Why is the relevance lost? In *The End of Accounting and the Path Forward for Investors and Managers*, pp. 77–80. Hoboken, NJ: John Wiley & Sons. https://doi.org/10.1002/9781119270041.part2

Lev, B. and Gu, F. (2016). Worse than at first sight. In *The End of Accounting and the Path Forward for Investors and Managers*, pp. 146–162. Hoboken, NJ: John Wiley & Sons. https://doi.org/10.1002/9781119270041.ch13

Liaw, K.T. (2012). Equity underwriting and IPOs. In *The Business of Investment Banking: A Comprehensive Review*, 3e, pp. 117–140. Hoboken, NJ: John Wiley & Sons. https://doi.org/10.1002/9781119202332.ch8

Liaw, K.T. (2012). Introduction to investment banking: how the financial crisis and reforms changed the industry. In *The Business of Investment Banking: A Comprehensive Review*, 3e, pp. 1–7. Hoboken, NJ: John Wiley & Sons. https://doi.org/10.1002/9781119202332.ch1

Liaw, K.T. (2012). Investment banking strategies: how they compete and profit. In *The Business of Investment Banking: A Comprehensive Review*, 3e, pp. 37–47. Hoboken, NJ: John Wiley & Sons. https://doi.org/10.1002/9781119202332.ch4

Liaw, K.T. (2012). New investment banking structure: financial holding companies, full-service, and boutique investment banks. In *The Business of Investment Banking: A Comprehensive Review*, 3e, pp. 9–20. Hoboken, NJ: John Wiley & Sons. https://doi.org/10.1002/9781119202332.ch2

Liaw, K.T. (2012). The BRICs: investment banking in Brazil, Russia, and India. In *The Business of Investment Banking: A Comprehensive Review*, 3e, pp. 323–338. Hoboken, NJ: John Wiley & Sons. https://doi.org/10.1002/9781119202332.ch17

Liaw, K.T. (2012). The structure of investment banks: divisions and services. In *The Business of Investment Banking: A Comprehensive Review*, 3e, pp. 21–36. Hoboken, NJ: John Wiley & Sons. https://doi.org/10.1002/9781119202332.ch3

Marks, K.H., Nall, M.R., Blees, C.W., and Stewart, T.A. (2022). Cross-border considerations. In *Middle Market M&A: Handbook for Advisors, Investors, and Business Owners*, 2e, pp. 359–376. Hoboken, NJ: John Wiley & Sons. https://onlinelibrary.wiley.com/doi/10.1002/9781119828150.ch20

Marks, K.H., Slee, R.T., Blees, C.W., and Nall, M.R. (2012). A global perspective. In *Middle Market M&A: Handbook for Investment Banking and Business Consulting*, pp. 51–70. Hoboken, NJ: John Wiley & Sons. https://doi.org/10.1002/9781119200659.ch4

Mawhinney, M. (2001). Consultant case studies. In *International Construction*, pp. 153–163. Oxford: Blackwell Science. https://doi.org/10.1002/9780470690628.ch7

Mellen, C.M. and Evans, F.C. (2012). Cross-border M&A. In *Valuation for M&A: Building Value in Private Companies*, 2e, pp. 313–327. Hoboken, NJ: John Wiley & Sons. https://doi.org/10.1002/9781119200154.ch19

Mellen, C.M. and Evans, F.C. (2018). Cross-border M&A. In *Valuation for M&A: Building and Measuring Private Company Value*, 3e, pp. 381–395. Hoboken, NJ: John Wiley & Sons. https://doi.org /10.1002/9781119437413.ch20

Muñoz-García, F.-J., López-Hernández, M,-Á, and Domínguez-Delgado, R. (2021). Written news search engines and retrieval systems of the databases of Spanish digital newspapers. *Proceedings of the Association for Information Science and Technology* 58(1): 795–797. https://doi.org /10.1002/pra2.565

Pan, S., Wang, L., Wang, K. et al. (2012). A knowledge engineering framework for identifying key impact factors from safety-related accident cases. *Systems Research and Behavioral Science* 31(3): 383–397. https://doi .org/10.1002/sres.2278

Pignataro, P. (2017). Introduction to investment banking. In *The Technical Interview Guide to Investment Banking*, p. 1. Hoboken, NJ: John Wiley & Sons. https://doi.org/10.1002/9781119161554.part1

Roberts, D.J. (2012). Investment banking representation on the buy side: some further thoughts for buyers. In *Mergers and Acquisitions: An Insider's Guide to the Purchase and Sale of Middle Market Business Interests*, pp. 165–176. Hoboken, NJ: John Wiley & Sons. https://doi .org/10.1002/9781119197386.ch14

Roberts, D.J. (2012). The middle market is different! In *Mergers and Acquisitions: An Insider's Guide to the Purchase and Sale of Middle Market Business Interests*, pp. 1–9. Hoboken, NJ: John Wiley & Sons. https:// doi.org/10.1002/9781119197386.ch1

Rodenas, M.A. and Jackson, M.C. (2021) Lessons for systems engineering from the Segura River Reclamation project: a critical systems thinking analysis. *Systems Research and Behavioral Science* 38(3): 368–376. https:// doi.org/10.1002/sres.2789

Rolland, G. (2012). Investment banking. In *Market Players: A Guide to the Institutions in Today's Financial Markets*, pp. 27–47. Chichester: John Wiley & Sons. https://doi.org/10.1002/9781119205845.ch3

Rousseau, D. (2019). A vision for advancing systems science as a foundation for the systems engineering and systems practice of the future. *Systems Research and Behavioral Science* 36(5): 621–634. https://doi .org/10.1002/sres.2629

Rousseau, D. (2020). The theoretical foundation(s) for systems engineering? Response to Yearworth. *Systems Research and Behavioral Science* 37(1): 188–191. https://doi.org/10.1002/sres.2671

Saleh, I. and Metwali, H. (2017). Dealing with political and cultural crisis in a troubled Middle East region. In *Culture and Crisis Communication: Cases from Nonwestern Perspectives*, ed. A.M. George and K. Kwansah-Aidoo, pp. 19–39. Piscataway, NJ: IEEE Press. https://doi.org/10.1002 /9781119081708.ch2

Saling, K.G. and White Jr., K.P. (2013). Integrating probabilistic design and rare-event simulation into the requirements engineering process for high-reliability systems. *International Transactions in Operational Research* 20(4): 515–531. https://doi.org/10.1111/itor.12023

Sanders, K. (2022). Mustafa Chehabbedine ARB. In *Voices of Design Leadership: Insights from Top Collaborative Design Firms*, pp. 61–71. Hoboken, NJ: John Wiley & Sons. https://doi.org/10.1002/9781119847359.ch6

Schindel, W.D. (2019). Identifying phenomenological foundations of systems engineering and systems science. *Systems Research and Behavioral Science* 36(5): 635–647. https://doi.org/10.1002/sres.2620

Schuster, C. and Fabiano, P. (2012). Doing business in Latin America: lessons learned and best practices for the protection of foreign investors. In *Risk Management in Finance: Six Sigma and Other Next-Generation Techniques*, ed. A. Tarantino and D. Cernauskas, pp. 75–85. Hoboken, NJ: John Wiley & Sons. https://doi.org/10.1002/9781119197812.ch8

Sollish, F. and Semanik, J. (2012). Logistics. In *The Procurement and Supply Chain Manager's Desk Reference*, pp. 327–349. Hoboken, NJ: John Wiley & Sons. https://doi.org/10.1002/9781119205098.ch19

Sueyoshi,T. and Goto, M. (2018). World energy. In *Environmental Assessment on Energy and Sustainability by Data Envelopment Analysis*, pp. 281–304. Hoboken, NJ: John Wiley & Sons. https://doi.org/10.1002/9781118979259.ch13

Tarantino, A. (2012). Financial risk management in Asia. In *Risk Management in Finance: Six Sigma and Other Next-Generation Techniques*, ed. A. Tarantino and D. Cernauskas, pp. 61–74. Hoboken, NJ: John Wiley & Sons. https://doi.org/10.1002/9781119197812.ch7

Wells, J.T. (2018). Latin America and the Caribbean. *International Fraud Handbook: Prevention and Detection*, pp. 265–285. Hoboken, NJ: John Wiley & Sons. https://doi.org/10.1002/9781119448709.ch10

World Bank (2023). World Bank Data, Access 2023. https://www.worldbank.org/en/home

Yearworth, M. (2020). The theoretical foundation(s) for systems engineering? *Systems Research and Behavioral Science* 37(1): 184–187. https://doi.org/10.1002/sres.2667

Yearworth, M. and Edwards, G. (2014). On the desirability of integrating research methods into overall systems approaches in the training of engineers: analysis using SSM. *Systems Research and Behavioral Science*. 31(1): 47–66. https://doi.org/10.1002/sres.2167

Index